Elementary Theory of Analytic Functions of One or Several Complex Variables

Henri Cartan

DOVER PUBLICATIONS, INC.
New York

Bibliographical Note

This Dover edition, first published in 1995, is an unabridged and unaltered republication of the 1973 second printing of the work first published by the Addison-Wesley Publishing Company, Reading, Massachusetts, in 1963 in the *ADIWES International Series in Mathematics*. It is a translation of a work originally published by Hermann, Paris, under the title *Théorie Élémentaire Des Fonctions Analytiques D'Une Ou Plusieurs Variables Complexes*.

Library of Congress Cataloging-in-Publication Data

Cartan, Henri Paul, 1904–
[Théorie élémentaire des fonctions analytiques d'une ou plusieurs variables complexes. English]
Elementary theory of analytic functions of one or several complex variables / Henri Cartan.
p. cm.
"An unabridged and unaltered republication of the 1973 second printing of the work first published by the Addison-Wesley Publishing Company, Reading, Massachusetts, in 1963 in the ADIWES international series in mathematics"—T.p. verso.
Includes indexes.
ISBN 0-486-68543-8 (pbk.)
1. Analytic functions. I. Title.
QA331.7.C3813 1995
515'.9—dc20 95-13507
CIP

Manufactured in the United States of America
Dover Publications, Inc., 31 East 2nd Street, Mineola, N.Y. 11501

TABLE OF CONTENTS

Preface

The present volume contains the substance, with some additions, of a course of lectures given at the Faculty of Science in Paris for the requirements of the *licence d'enseignement* during the academic sessions 1957-1958, 1958-1959 and 1959-1960. It is basically concerned with the theory of analytic functions of a complex variable. The case of analytic functions of several real or complex variables is, however, touched on in chapter IV if only to give an insight into the harmonic functions of two real variables as analytic functions and to permit the treatment in chapter VII of the existence theorem for the solutions of differential systems in cases where the data is analytic.

The subject matter of this book covers that part of the " Mathematics II " certificate syllabus given to analytic functions. This same subject matter was already included in the " Differential and integral calculus " certificate of the old *licence.*

As the syllabuses of certificates for the *licence* are not fixed in detail, the teacher usually enjoys a considerable degree of freedom in choosing the subject matter of his course. This freedom is mainly limited by tradition and, in the case of analytic functions of a complex variable, the tradition in France is fairly well established. It will therefore perhaps be useful to indicate here to what extent I have departed from this tradition. In the first place I decided to begin by offering not Cauchy's point of view (differentiable functions and Cauchy's integral) but the Weierstrass point of view, i.e. the theory of convergent power series (chapter I). This is itself preceded by a brief account of formal operations on power series, i.e. what is called nowadays the theory of formal series. I have also made something of an innovation by devoting two paragraphs of chapter VI to a systematic though very elementary exposition of the theory of abstract complex manifolds of one complex dimension. What is referred to here as a complex manifold is simply what used to be called a Riemann surface and is often still given that name; for our part, we decided to keep the term Riemann surface for the double datum of a complex manifold and a holomorphic mapping of this manifold into the complex plane (or, more generally, into another complex manifold). In this way a distinction is made between the two ideas with a clarity unattainable with orthodox terminology. With a subject as well established as the theory of analytic functions of a complex variable, which has been in the past the subject of so many treatises and still is in all countries, there could be no question of laying claim to originality. If the present treatise differs in any way from its forerunners in France, it does so perhaps because it conforms to a recent practice which is becoming increasingly prevalent: a mathematical text must contain precise statements of propositions or theorems — statements which are adequate in themselves and to which reference can be made at all times. With a very few exceptions which are clearly indicated, complete proofs are given of all the statements in the text. The somewhat ticklish problems of plane topology in relation to Cauchy's integral and the discussion of many-valued functions are approached quite openly in chapter II. Here again it was thought that a few precise statements were preferable to vague intuitions and hazy ideas. On these problems of plane topology, I drew my inspiration from the excellent book by L. Ahlfors (Complex Analysis), without however conforming completely with the points of view he develops. The basic concepts of general Topology are assumed to be familiar to the reader and are employed frequently in the present work; in fact this course is addressed to students of 'Mathematics II' who are expected to have already studied the 'Mathematics I' syllabus.

I express my hearty thanks to Monsieur Reiji Takahashi, who are from experience gained in directing the practical work of students, has consented to supplement the various chapters of this book with exersices and problems. It is hoped that the reader will thus be in a position to make sure that he has understood and assimilated the theoretical ideas set out in the text.

HENRI CARTAN

Die (Drôme), August 4th, 1960

CHAPTER I

Power Series in One Variable

1. Formal Power Series

1. Algebra of polynomials

Let K be a commutative field. We consider the formal polynomials in one symbol (or 'indeterminate') X with coefficients in K (for the moment we do not give a value to X). The laws of addition of two polynomials and of multiplication of a polynomial by a 'scalar' makes the set K[X] of polynomials into a *vector space* over K with the infinite base

$$1, X, \ldots, X^n, \ldots$$

Each polynomial is a finite linear combination of the X^n with coefficients in K and we write it $\sum_{n\geqslant 0} a_n X^n$, where it is understood that only a finite number of the coefficients a_n are non-zero in the infinite sequence of these coefficients. The multiplication table

$$X^p . X^q = X^{p+q}$$

defines a multiplication in K[X]; the product

$$\left(\sum_p a_p X^p\right) . \left(\sum_q b_q X^q\right)$$

is $\sum_n c_n X^n$, where

$$c_n = \sum_{p+q=n} a_p b_q. \tag{1.1}$$

This multiplication is commutative and associative. It is bilinear in the sense that

$$\begin{cases} (P_1 + P_2) \cdot Q = P_1 Q + P_2 Q \\ (\lambda P) \cdot Q = \lambda \ (PQ) \end{cases} \tag{1.2}$$

for all polynomials P, P_1, P_2, Q and all scalars λ. It admits as unit element (denoted by 1) the polynomial $\sum_{n\geqslant 0} a_n X^n$ such that $a_0 = 1$ and $a_n = 0$ for $n > 0$. We express all these properties by saying that K[X], provided with its vector space structure and its multiplication, is a *commutative algebra* with a unit element over the field K; it is, in particular, a commutative ring with a unit element.

2. The algebra of formal series

A formal power series in X is a formal expression $\sum_{n\geqslant 0} a_n X^n$, where this time we no longer require that only a finite number of the coefficients a_n are non-zero. We define the sum of two formal series by

$$\Big(\sum_{n\geqslant 0} a_n X^n\Big) + \Big(\sum_{n\geqslant 0} b_n X^n\Big) = \sum_{n\geqslant 0} c_n X^n, \qquad \text{where} \qquad c_n = a_n + b_n,$$

and the product of a formal series with a scalar by

$$\lambda\Big(\sum_{n\geqslant 0} a_n X^n\Big) = \sum_{n\geqslant 0} (\lambda a_n) X^n.$$

The set K[[X]] of formal series then forms a vector space over K. The neutral element of the addition is denoted by 0; it is the formal series with all its coefficients zero.

The product of two formal series is defined by the formula (1.1), which still has a meaning because the sum on the right hand side is over a finite number of terms. The multiplication is still commutative, associative and bilinear with respect to the vector structure. Thus K[[X]] is an algebra over the field K with a unit element (denoted by 1), which is the series $\sum_{n\geqslant 0} a_n X^n$ such that $a_0 = 1$ and $a_n = 0$ for $n > 0$.

The algebra K[X] is identified with a subalgebra of K[[X]], the subalgebra of formal series whose coefficients are all zero except for a finite number of them.

3. The order of a formal series

Denote $\sum_{n\geqslant 0} a_n X^n$ by S(X), or, more briefly, by S. The *order* $\omega(S)$ of this series is an integer which is only defined when $S \neq 0$; it is the smallest n such that $a_n \neq 0$. We say that a formal series S has order $\geqslant k$ if it is 0 or if $\omega(S) \geqslant k$. By *abus de langage*, we write $\omega(S) \geqslant k$ even when $S = 0$ although $\omega(S)$ is not defined in this case.

Note. We can make the convention that $\omega(0) = +\infty$. The S such that $\omega(S) \geqslant k$ (for a given integer k) are simply the series $\sum_{n \geqslant 0} a_n X^n$ such that $a_n = 0$ for $n < k$. They form a vector subspace of K[[X]].

Definition. A family $(S_i(X))_{i \in I}$, where I denotes a set of indices, is said to be *summable* if, for any integer k, $\omega(S_i) \geqslant k$ for all but a finite number of the indices i. By definition, the *sum* of a summable family of formal series

$$S_i(X) = \sum_{n \geqslant 0} a_{n,i} X^n$$

is the series

$$S(X) = \sum_{n \geqslant 0} a_n X^n,$$

where, for each n, $a_n = \sum_i a_{n,i}$. This makes sense because, for fixed n, all but a finite number of the $a_{n,i}$ are zero by hypothesis. The operation of addition of formal series which form summable families generalizes the finite addition of the vector structure of K[[X]]. The generalized addition is commutative and associative in a sense which the reader should specify.

The formal notation $\sum_{n \geqslant 0} a_n X^n$ can then be justified by what follows. Let a *monomial* of degree p be a formal series $\sum_{n \geqslant 0} a_n X^n$ such that $a_n = 0$ for $n \neq p$ and let $a_p X^p$ denote such a monomial. The family of monomials $(a_n X^n)_{n \in N}$ (N being the set of integers $\geqslant 0$) is obviously summable, and its sum is simply the formal series $\sum_{n \geqslant 0} a_n X^n$.

Note. The product of two formal series

$$\left(\sum_p a_p X^p\right) \cdot \left(\sum_q b_q X^q\right)$$

is merely the sum of the summable family formed by all the products

$$(a_p X^p) \cdot (b_q X^q) = (a_p b_q) X^{p+q}$$

of a monomial of the first series by one of the second.

PROPOSITION 3.1. *The ring* K[[X]] *is an integral domain* (this means that $S \neq 0$ and $T \neq 0$ imply $ST \neq 0$).

Proof. Suppose that $S(X) = \sum_p a_p X^p$ and $T(X) = \sum_q b_q X^q$ are non-zero. Let $p = \omega(S)$ and $q = \omega(T)$, let

$$S(X) \cdot T(X) = \sum_n c_n X^n;$$

obviously $c_n = 0$ for $n < p + q$ and $c_{p+q} = a_p b_q$. Since K is a field and since $a_p \neq 0$, $b_q \neq 0$, we have that $c_{p+q} \neq 0$, so ST is not zero.

What is more, we have proved that

$$(3.1) \qquad \omega(S \cdot T) = \omega(S) + \omega(T) \qquad \text{for} \qquad S \neq 0 \quad \text{and} \quad T \neq 0.$$

Note. One can consider formal series with coefficients in a commutative ring A with a unit element which is not necessarily a field K; the above proof then establishes that, *if* A *is an integral domain, then so is* A[[X]].

4. Substitution of a formal series in another

Consider two formal series

$$S(X) = \sum_{n \geq 0} a_n X^n, \qquad T(Y) = \sum_{p \geq 0} b_p Y^p.$$

It is essential also to assume that $b_0 = 0$, in other words that $\omega(T) \geq 1$. To each monomial $a_n X^n$ associate the formal series $a_n(T(Y))^n$, which has a meaning because the formal series in Y form an algebra. Since $b_0 = 0$, the order of $a_n(T(Y))^n$ is $\geq n$; thus the family of the $a_n(T(Y))^n$ (as n takes the values 0, 1, ...) is *summable*, and we can consider the formal series

$$(4.1) \qquad \sum_{n \geq 0} a_n(T(Y))^n,$$

in which we regroup the powers of Y. This formal series in Y is said to be obtained *by substitution* of T(Y) for X in S(X); we denote it by S(T(Y)), or $S \circ T$ without specifying the indeterminate Y. The reader will verify the relations :

$$(4.2) \qquad \begin{cases} (S_1 + S_2) \circ T = S_1 \circ T + S_2 \circ T, \\ (S_1 S_2) \circ T = (S_1 \circ T)(S_2 \circ T), \qquad 1 \circ T = 1. \end{cases}$$

But, note carefully that $S \circ (T_1 + T_2)$ is not, in general, equal to

$$S \circ T_1 + S \circ T_2.$$

The relations (4.2) express that, for given T (of order ≥ 1), the mapping $S \to S \circ T$ is a *homomorphism* of the ring K[[X]] in the ring K[[Y]] which transforms the unit element 1 into 1.

Note. If we substitute 0 in $S(X) = \sum_{n \geq 0} a_n X^n$, we find that the formal series reduces to its 'constant term' a_0.

If we have a summable family of formal series S_i and if $\omega(T) \geqslant 1$, then the family $S_i \circ T$ is summable and

$$\left(\sum_i S_i\right) \circ T = \sum (S_i \circ T), \tag{4.3}$$

which generalizes the first of the relations (4. 2). For, let

$$S_i(X) = \sum_{n \geqslant 0} a_{n,i} X^n;$$

we have

$$\sum_i S_i(X) = \sum_{n \geqslant 0} \left(\sum_i a_{n,i}\right) X^n,$$

whence

$$\left(\sum_i S_i\right) \circ T = \sum_{n \geqslant 0} \left(\sum_i a_{n,i}\right) (T(Y))^n, \tag{4.4}$$

while

$$\sum S_i \circ T = \sum_i \left(\sum_{n \geqslant 0} a_{n,i} (T(Y))^n\right). \tag{4.5}$$

To prove the equality of the right hand sides of (4. 4) and (4. 5), we observe that the coefficient of a given power Y^p in each of them involves only a finite number of the coefficients $a_{n,i}$ and we apply the associativity law of (finite) addition in the field K.

PROPOSITION 4. 1. *The relation*

$$(S \circ T) \circ U = S \circ (T \circ U) \tag{4.6}$$

holds whenever $\omega(T) \geqslant 1$, $\omega(U) \geqslant 1$ (associativity of substitution).

Proof. Both sides of (4. 6) are defined. In the case when S is a monomial, they are equal because

$$T^n \circ U = (T \circ U)^n \tag{4.7}$$

which follows by induction on n from the second relation in (4. 2).

The general case of (4. 6) follows by considering the series S as the (infinite) sum of its monomials $a_n X^n$; by definition,

$$S \circ T = \sum_{n \geqslant 0} a_n T^n,$$

and, from (4. 3),

$$(S \circ T) \circ U = \sum_{n \geqslant 0} a_n (T^n \circ U),$$

which, by (4. 7), is equal to

$$\sum_{n\geqslant 0} a_n(T \circ U)^n = S \circ (T \circ U).$$

This completes the proof.

5. Algebraic inverse of a formal series

In the ring K[[Y]], the identity

$$(5.1) \qquad (1 - Y)(1 + Y + \cdots + Y^n + \cdots) = 1$$

can easily be verified. Hence the series $1 - Y$ has an inverse in K[[Y]]

Proposition 5. 1. *For* $S(X) = \sum_n a_n X^n$ *to have an inverse element for the multiplication of* K[[X]], *it is necessary and sufficient that* $a_0 \neq 0$, *i.e. that* $S(0) \neq 0$.

Proof. The condition is necessary because, if

$$T(X) = \sum_n b_n X^n \qquad \text{and if} \qquad S(X)T(X) = 1,$$

then $a_0 b_0 = 1$ and so $a_0 \neq 0$. Conversely, suppose that $a_0 \neq 0$; we shall show that $(a_0)^{-1}S(X) = S_1(X)$ has an inverse $T_1(X)$, whence it follows that $(a_0)^{-1}T_1(X)$ is the inverse of $S(X)$. Now

$$S_1(X) = 1 - U(X) \qquad \text{with} \qquad \omega(U) \geqslant 1,$$

and we can substitute $U(X)$ for Y in the relation (5. 1), from which it follows that $1 - U(X)$ has an inverse. The proposition is proved.

Note. By considering the algebra of polynomials K[X] imbedded in the algebra of formal series K[[X]], it will be seen that any polynomial $Q(X)$ such that $Q(0) \neq 0$ has an inverse in the ring K[[X]]; this ring then contains all the quotients $P(X)/Q(X)$, where P and Q are polynomials and where $Q(0) \neq 0$.

6. Formal derivative of a series

Let $S(X) = \sum_n a_n X^n$; by definition, the *derived* series $S'(X)$ is given by the formula

$$(6.1) \qquad S'(X) = \sum_{n\geqslant 0} n a_n X^{n-1}.$$

It can also be written $\frac{dS}{dX}$, or $\frac{d}{dX}S$. The derivative of a (finite or infinite) sum is equal to the sum of its derivatives. The mapping $S \to S'$ is a linear mapping of K[[X]] into itself. Moreover, the derivative of the product of two formal series is given by the formula

$$\frac{d}{dX}(ST) = \frac{dS}{dX}T + S\frac{dT}{dX}. \tag{6. 2}$$

For, it is sufficient to verify this formula in the particular case when S and T are monomials, and it is clearly true then.

If $S(0) \neq 0$, let T be the inverse of S (c.f. nº. 5). The formula (6. 2) gives

$$\frac{d}{dX}\left(\frac{1}{S}\right) = -\frac{1}{S^2}\frac{dS}{dX}. \tag{6. 3}$$

Higher derivatives of a formal series are defined by induction. If $S(X) = \sum a_n X^n$, its derivative of order n is

$$S^{(n)}(X) = n!\, a_n + \text{terms of order} \geqslant 1.$$

Hence,

$$S^{(n)}(0) = n!\, a_n, \tag{6. 4}$$

where $S^{(n)}(0)$ means the result of substituting the series 0 for the indeterminate X in $S^{(n)}(X)$.

7. Compositional inverse series

The series I(X) defined by $I(X) = X$ is a *neutral element* for the composition of formal series :

$$S \circ I = S = I \circ S.$$

PROPOSITION 7. 1. *Given a formal series* S, *a necessary and sufficient condition for there to exist a formal series* T *such that*

$$T(0) = 0, \qquad S \circ T = I \tag{7. 1}$$

is that

$$S(0) = 0, \qquad S'(0) \neq 0. \tag{7. 2}$$

In this case, T *is unique, and* $T \circ S = I$: *in other words* T *is the inverse of* S *for the law of composition* $\circ$.

Proof. Let $S(X) = \sum_{n \geqslant 0} a_n X^n$, $T(Y) = \sum_{n \geqslant 1} b_n Y^n$. If

$$S(T(Y)) = Y, \tag{7. 3}$$

then equating the first two terms gives

$$a_0 = 0, \qquad a_1 b_1 = 1. \tag{7.4}$$

Hence the conditions (7. 2) are necessary.

Suppose that they are satisfied; we write down the condition that the coefficient of Y^n is zero in the left hand side of (7. 3). This coefficient is the same as the coefficient of Y^n in

$$a_1 T(Y) + a_2 (T(Y))^2 + \cdots + a_n (T(Y))^n,$$

which gives the relation

$$a_1 b_n + P_n(a_2, \ldots, a_n, b_1, \ldots, b_{n-1}) = 0, \tag{7.5}$$

where P_n is a known polynomial with non-negative integral coefficients and is linear in $a_2, \ldots, a_n$. Since $a_1 \neq 0$, the second equation (7. 4) determines b_1; then, for $n \geqslant 2$, b_n can be calculated by induction on n from (7. 5). Thus we have the existence and uniqueness of the formal series $T(Y)$. The series thus obtained satisfies $T(0) = 0$ and $T'(0) \neq 0$, and so the result that we have just proved for S can be applied to T, giving a formal series S_1 such that

$$S_1(0) = 0, \qquad T \circ S_1 = I.$$

This implies that

$$S_1 = I \circ S_1 = (S \circ T) \circ S_1 = S \circ (T \circ S_1) = S \circ I = S.$$

Hence S_1 is none other than S and, indeed, $T \circ S = I$, which completes the proof.

Remark. Since $S(T(Y)) = Y$ and $T(S(X)) = X$, we can say that the 'formal transformations'

$$Y = S(X), \qquad X = T(Y)$$

are inverse to one another; thus we call T the 'inverse formal series' of the series S.

Proposition 7. 1 is an 'implicit function theorem' for formal functions.

2. Convergent power series

1. The complex field

From now on, the field K will be either **R** or **C**, where **R** denotes the field of real numbers and **C** the field of complex numbers.

Recall that a complex number $z = x + iy$ (x and y real) is represented by a point on the plane $\mathbf{R}^2$ whose coordinates are x and y. If we associate with each complex number $z = x + iy$ its 'conjugate' $\bar{z} = x - iy$, we define an automorphism $z \to \bar{z}$ of the field **C**, since

$$\overline{z + z'} = \bar{z} + \bar{z}', \qquad \overline{zz'} = \bar{z}\bar{z}'.$$

The conjugate of $\bar{z}$ is z; in other words, the transformation $z \to \bar{z}$ is *involutive*, i.e. is equal to its inverse transformation.

The *norm*, *absolute value*, or *modulus* $|z|$ of a complex number z is defined by

$$|z| = (z \cdot \bar{z})^{1/2}.$$

It has the following properties :

$$|z + z'| \leqslant |z| + |z'|, \qquad |zz'| = |z| . |z'|, \qquad |1| = 1.$$

The norm $|z|$ is always $\geqslant 0$ and is zero only when $z = 0$. This norm enables us to define a *distance* in the field **C** : the distance between z and z' is $|z - z'|$, which is precisely the euclidean distance in the plane $\mathbf{R}^2$. The space **C** is a *complete* space for this distance function, which means that the *Cauchy criterion* is valid : for a sequence of points $z_n \in \mathbf{C}$ to have a limit, it is necessary and sufficient that

$$\lim_{\substack{m \to \infty \\ n \to \infty}} |z_m - z_n| = 0.$$

The Cauchy criterion gives the following well-known theorem : if a series $\sum_n u_n$ of complex numbers is such that $\sum_n |u_n| < +\infty$, then the series converges (we say that the series is absolutely convergent). Moreover,

$$\left|\sum_n u_n\right| \leqslant \sum_n |u_n|.$$

We shall always identify **R** with a sub-field of **C**, i. e. the sub-field formed by the z such that $\bar{z} = z$. The norm induces a norm on **R**, which is merely the absolute value of the real number. **R** is complete. The norm of the field **C** (or **R**) plays an essential role in what follows.

We define

$$\operatorname{Re}(z) = \frac{1}{2}(z + \bar{z}) \qquad \text{and} \qquad \operatorname{Im}(z) = \frac{1}{2i}(z - \bar{z})$$

the 'real part' and the 'imaginary coefficient' of $z \in \mathbf{C}$.

2. Revision of the theory of convergence of series of functions

(For a more complete account of this theory, the reader is referred to Cours de Mathématiques I of J. Dixmier : Cours de l'A.C.E.S., Topologie, chapter VI, § 9.)

Consider functions defined on a set E taking real, or complex, values (or one could consider the more general case when the functions take values in a complete normed vector space; cf. *loc. cit.*). For each function u, we write

$$\|u\| = \sup_{x \in E} |u(x)|,$$

which is a number $\geqslant 0$, or may be infinite. Evidently,

$$\|u + v\| \leqslant \|u\| + \|v\|, \qquad \|\lambda u\| = |\lambda| . \|u\|$$

for any scalar λ, when $\|u\| < +\infty$: in other words, $\|u\|$ is a *norm* on the vector space of functions u such that $\|u\| < +\infty$.

We say that a series of functions u_n is *normally convergent* if the series of norms $\sum_n \|u_n\|$ is a convergent series of positive terms, in other words, if $\sum_n \|u_n\| < +\infty$. This implies that, for each $x \in E$, the series $\sum_n |u_n(x)|$ is convergent, and so the series $\sum_n u_n(x)$ is absolutely convergent; moreover, if $v(x)$ is the sum of this last series,

$$\|v\| \leqslant \sum_n \|u_n\|, \qquad \lim_{p \to \infty} \|v - \sum_{n=0}^{p} u_n\| = 0.$$

The latter relation expresses that the partial sums $\sum_{n=0}^{p} u_n$ converge *uniformly* to v as P tends to infinitiy. Thus, *a normally convergent series is uniformly convergent*. If A is a subset of E, the series whose general term is u_n is said to converge normally for $x \in A$ if the series of functions

$$u'_n = u_n | A \quad \text{(restriction of } u_n \text{ to A)}$$

is normally convergent. This is the same as saying that we can bound each $|u_n(x)|$ on A above by a constant $\varepsilon_n \geqslant 0$ in such a way that the series $\sum_n \varepsilon_n$ is convergent. Recall that the limit of a uniformly convergent sequence of *continuous* functions (on a topological space E) is continuous. In particular, *the sum of a normally convergent series of continuous functions is continuous*. An important consequence of this is :

Proposition 1.2. *Suppose that, for each* n, $\lim_{x \to x_0} u_n(x)$ *exists and takes the value* a_n. *Then, if the series* $\sum_n u_n$ *is normally convergent, the series* $\sum_n a_n$ *is convergent and*

$$\sum_n a_n = \lim_{x \to x_0} \left(\sum_n u_n(x) \right)$$

(changing the order of the summation and the limiting process).

All these results extend to multiple series and, more generally, to summable families of functions (cf. the above-mentioned course by Dixmier).

3. Radius of convergence of a power series

All the power series to be considered will have coefficients in either the field **R**, or the field **C**.

Note however that what follows remains valid in the more general case when coefficients are in any field with a complete, non-discrete, valuation, that is, a field K with a mapping $x \to |x|$ of K into the set of real numbers $\geqslant 0$ such that

$$\begin{cases} |x+y| \leqslant |x| + |y|, \qquad |xy| = |x|.|y|, \\ (|x| = 0) \Longleftrightarrow (x = 0), \end{cases}$$

and such that there exists some $x \neq 0$ with $|x| \neq 1$.

Let $S(X) = \sum_{n \geqslant 0} a_n X^n$ be a formal series with coefficients in **R** or **C**. We propose to substitute an element z of the field for the indeterminate X and thus to obtain a 'value' $S(z)$ of the series, which will be an element of the field; but this substitution is not possible unless the series $\sum_{n \geqslant 0} a_n z^n$ is convergent. In fact, we shall limit ourselves to the case when it is *absolutely convergent.*

To be precise, we introduce a real variable $r \geqslant 0$ and consider the series of positive (or zero) terms

$$\sum_{n \geqslant 0} |a_n| r^n,$$

called the *associated series* of $S(X)$. Its sum is a well-defined number $\geqslant 0$, which may be infinity. The set of $r \geqslant 0$ for which

$$\sum_{n \geqslant 0} |a_n| r^n < +\infty$$

is clearly an interval of the half line $\mathbf{R}^+$, and this interval is non-empty since the series converges for $r = 0$. The interval can either be open or closed on the right, it can be finite or infinite, or it can reduce at the single point 0. In all cases, let ρ be the least upper bound of the interval, so ρ is a number $\geqslant 0$, finite, infinite, or zero; it is called the *radius of convergence* of the formal power series $\sum_{n \geqslant 0} a_n X^n$. The set of z such that $|z| < \rho$ is called the *disc of convergence* of the power series; it is an open set and it is empty if $\rho = 0$. It is an ordinary disc when the field of coefficients is the complex field **C**.

Proposition 3. 1.

a) *For any $r < \rho$, the series $\sum_{n \geqslant 0} a_n z^n$ converges normally for $|z| \leqslant r$. In particular, the series converges absolutely for each z such that $|z| < \rho$;*

b) *the series* $\sum_{n \geqslant 0} a_n z^n$ *diverges for* $|z| > \rho$. (We say nothing about the case when $|z| = \rho$.)

Proof. Proposition 3.1 follows from

ABEL'S LEMMA. *Let* r *and* r_0 *be real numbers such that* $0 < r < r_0$. *If there exists a finite number* $M > 0$ *such that*

$$|a_n|(r_0)^n \leqslant M \qquad \textit{for any integer } n \geqslant 0,$$

then the series $\sum_{n \geqslant 0} a_n z^n$ *converges normally for* $|z| \leqslant r$.

For, $|a_n z^n| \leqslant |a_n| r^n \leqslant M(r/r_0)^n$, and $\varepsilon_n = M(r/r_0)^n$ is the general term of a convergent series — a geometric series with common ratio $r/r_0 < 1$. We now prove statement a) of proposition 3.1 : if $r < \rho$, choose r_0 such that $r < r_0 < \rho$; since $\sum_{n \geqslant 0} |a_n|(r_0)^n$ converges, its general term is bounded above by a fixed number M, and Abel's lemma ensures the normal convergence of $\sum_{n \geqslant 0} a_n z^n$ for $|z| \leqslant r$. Statement b) remains to be proved : if $|z| > \rho$, we can make $|a_n z^n|$ arbitrarily large by chosing the integer n suitably because, otherwise, Abels' lemma would give an r' with $\rho < r' < |z|$ such that the series $\sum_{n \geqslant 0} |a_n| r'^n$ were convergent and this would contradict the definition of ρ.

Formula for the radius of convergence (Hadamard) : we shall prove the formula

$$1/\rho = \limsup_{n \to \infty} |a_n|^{1/n}. \tag{3.1}$$

Recall, first of all, the definition of the upper limit of a sequence of real numbers u_n:

$$\limsup_{n \to \infty} u_n = \lim_{p \to \infty} \left(\sup_{n \geqslant p} u_n\right).$$

To prove (3. 1), we use a classical criterion of convergence : if v_n is a sequence of non-negative numbers such that $\limsup_{n \to \infty} (v_n)^{1/n} < 1$, then $\sum_n v_n < +\infty$; moreover, if they are such that $\limsup_{n \to \infty} (v_n)^{1/n} > 1$, then $\sum_n v_n = +\infty$ (this is " Cauchy's rule " and follows by comparing the series $\sum_n v_n$ with a geometric series).

Here we put $v_n = |a_n| r^n$ and find that

$$\limsup_{n \to \infty} (v_n)^{1/n} = r\left(\limsup_{n \to \infty} |a_n|^{1/n}\right),$$

and so the series $\sum_n |a_n| r^n$ converges for $1/r > \limsup_{n\to\infty} |a_n|^{1/n}$, and diverges for $1/r < \limsup_{n\to\infty} |a_n|^{1/n}$. This proves (3. 1).

Some examples. — The series $\sum_{n\geq 0} n! z^n$ has zero radius of convergence;
— the series $\sum_{n\geq 0} \frac{1}{n!} z^n$ has infinite radius of convergence;
— each of the series $\sum_{n\geq 0} z^n$, $\sum_{n>0} \frac{1}{n} z^n$, $\sum_{n>0} \frac{1}{n^2} z^n$ has radius of convergence equal to 1. It can be shown that they behave differently when $|z| = 1$.

4. Addition and multiplication of convergent power series.

Proposition 4. 1. *Let* A(X) *and* B(X) *be two formal power series whose radii of convergence are* $\geq \rho$. *Let*

$$S(X) = A(X) + B(X) \quad \text{and} \quad P(X) = A(X).B(X)$$

be their sum and product. Then :

a) *the series* S(X) *and* P(X) *have radius of convergence* $\geq \rho$;

b) *for* $|z| < \rho$, we have

$$(4.1) \qquad S(z) = A(z) + B(z), \qquad P(z) = A(z)B(z).$$

Proof. Let

$$A(X) = \sum_{n\geq 0} a_n X^n, \quad B(X) = \sum_{n\geq 0} b_n X^n, \quad S(X) = \sum_{n\geq 0} c_n X^n, \quad P(X) = \sum_{n\geq 0} d_n X^n,$$

and let

$$\gamma_n = |a_n| + |b_n|, \qquad \delta_n = \sum_{0\leq p\leq n} |a_p| . |b_{n-p}|.$$

We have $|c_n| \leq \gamma_n$, $|d_n| \leq \delta_n$. If $r < \rho$, the series $\sum_{n\geq 0} |a_n| r^n$ and $\sum_{n\geq 0} |b_n| r^n$ converge, thus

$$\sum_{n\geq 0} \gamma_n r^n = \Big(\sum_{n\geq 0} |a_n| r^n\Big) + \Big(\sum_{n\geq 0} |b_n| r^n\Big) < +\infty,$$
$$\sum_{n\geq 0} \delta_n r^n = \Big(\sum_{p\geq 0} |a_p| r^p\Big) . \Big(\sum_{q\geq 0} |b_q| r^q\Big) < +\infty.$$

It follows that the series $\sum_{n\geq 0} |c_n| r^n$ and $\sum_{n\geq 0} |d_n| r^n$ converge and therefore that any $r < \rho$ is less than or equal to the radius of convergence of each of the series S(X) and P(X). Thus both radii of convergence are $\geq \rho$. The two relations (4. 1) remain to be proved. The first is obvious, and the second is obtained by multiplying convergent series; to be precise, we recall this classical result :

PROPOSITION 4. 2. *Let* $\sum_{n\geqslant 0} u_n$ *and* $\sum_{n\geqslant 0} v_n$ *be two absolutely convergent series. If*

$$w_n = \sum_{0\leqslant p\leqslant n} u_p v_{n-p},$$

then the series $\sum_{n\geqslant 0} w_n$ *is absolutely convergent and its sum is equal to the product*

$$\left(\sum_{p\geqslant 0} u_p\right)\cdot\left(\sum_{q\geqslant 0} v_q\right)$$

Write $\alpha_p = \sum_{n\geqslant p} |u_n|$, $\beta_q = \sum_{n\geqslant q} |v_n|$; we have

$$\sum_{n\geqslant 0} |w_n| \leqslant \sum_{p\geqslant 0}\sum_{q\geqslant 0} |u_p|.|v_q| = \alpha_0\beta_0;$$

moreover, if $m \geqslant 2n$,

$$\sum_{k\leqslant m} w_k - \left(\sum_{k\leqslant n} u_k\right)\cdot\left(\sum_{k\leqslant n} v_k\right)$$

is less than a sum of terms $|u_p|.|v_q|$, where for each term, at least one of the integers p and q is $> n$; thus, this sum is less than $\alpha_0\beta_{n+1}+\beta_0\alpha_{n+1}$, which tends to zero as n tends to infinity. It follows that $\sum_{k\leqslant m} w_k$ tends to the product of the infinite sums $\sum_{n\geqslant 0} u_n$ and $\sum_{n\geqslant 0} v_n$.

5. SUBSTITUTION OF A CONVERGENT POWER SERIES IN ANOTHER

For two given formal power series S and T with $T(0) = 0$, we have defined the formal power series $S \circ T$ in paragraph 1, no. 4.

PROPOSITION 5.1. *Suppose* $T(X) = \sum_{n\geqslant 1} b_n X^n$. *If the radii of convergence* $\rho(S)$ *and* $\rho(T)$ *are* $\neq 0$, *then the radius of convergence of* $U = S \circ T$ *is also* $\neq 0$. *To be precise, there exists an* $r > 0$ *such that* $\sum_{n\geqslant 1} |b_n| r^n < \rho(S)$; *the radius of convergence of* U *is* $\geqslant r$, *and, for any* z *such that* $|z| \leqslant r$, *we have*

$$|T(z)| < \rho(S)$$

and

$$S(T(z)) = U(z). \tag{5.1}$$

Proof. Put $S(X) = \sum_{n\geqslant 0} a_n X^n$. For sufficiently small $r > 0$, $\sum_{n\geqslant 1} |b_n| r^n$ is finite since the radius of convergence of T is $\neq 0$. Thus, $\sum_{n\geqslant 1} |b_n| r^{n-1}$ is finite for sufficiently small $r > 0$, and, consequently,

$$\sum_{n\geqslant 1} |b_n| r^n = r.\left(\sum_{n\geqslant 1} |b_n| r^{n-1}\right)$$

tends to 0 when r tends to 0. There exists, then, an $r > 0$ such that $\sum_{n\geqslant 1} |b_n| r^n < \rho(S)$ as required. It follows that

$$\sum_{p\geqslant 0} |a_p| \left(\sum_{k\geqslant 1} |b_k| r^k\right)^p$$

is finite. However, this is a series $\sum_{n\geqslant 0} \gamma_n r^n$, and, if we put $U(x) = \sum_{n\geqslant 0} c_n X^n$, we clearly obtain $|c_n| \leqslant \gamma_n$. Thus $\sum_{n\geqslant 0} |c_n| r^n$ is finite and the radius of convergence of U is $\geqslant r$.

Relation (5. 1) remains to be proved. Put $S_n(X) = \sum_{0\leqslant k\leqslant n} a_k X^k$ and let $S_n \circ T = U_n$. For $|z| \leqslant r$, we have

$$U_n(z) = S_n(T(z)),$$

since the mapping $T \to T(z)$ is a ring homomorphism and S_n is a polynomial. Since the series S converges at the point $T(z)$, we have

$$S(T(z)) = \lim_n S_n(T(z)).$$

On the other hand, the coefficients of $U - U_n = (S - S_n) \circ T$ are bounded by those of

$$\sum_{p>n} |a_p| \left(\sum_{k\geqslant 1} |b_k| r^k\right)^p,$$

a series whose sum tends to 0 as $n \to +\infty$. It follows that, for $|z| \leqslant r$, $U(z) - U_n(z)$ tends to 0 as $n \to +\infty$. Finally, we have

$$U(z) = \lim_{n\to\infty} U_n(z) = \lim_{n\to\infty} S_n(T(z)) = S(T(z)) \qquad \text{for} \qquad |z| \leqslant r,$$

which establishes relation (5. 1) and completes the proof.

Interpretation of relation (5. 1) : suppose r satisfies the conditions of proposition 5. 1. Denote the function $z \to T(z)$ by $\tilde{T}$, defined for $|z| \leqslant r$, and similarly denote the functions defined by the series S and U by $\tilde{S}$ and $\tilde{U}$ respectively. The relation (5. 1) expresses that, for $|z| \leqslant r$, the composite function $\tilde{S} \circ \tilde{T}$ is defined and is equal to $\tilde{U}$. Thus the relation $U = S \circ T$ between formal series implies the relation $\tilde{U} = \tilde{S} \circ \tilde{T}$ if the radii of convergence of S and T are $\neq 0$ and *if we restrict ourselves to sufficiently small values of the variable z.*

6. Algebraic inverse of a convergent power series

We know (§ 1, proposition 5. 1) that, if $S(X) = \sum_{n \geqslant 0} a_n X^n$ with $a_0 \neq 0$, there exists a unique formal series $T(X)$ such that $S(X)T(X)$ is equal to 1.

Proposition 6. 1. *If the radius of convergence of* S *is* $\neq 0$, *then the radius of convergence of the series* T *such that* $ST = 1$ *is also* $\neq 0$.

Proof. Multiplying $S(X)$ by a suitable constant reduces the proposition to the special case when $a_0 = 1$. Put $S(X) = 1 - U(X)$ so that $U(0) = 0$. The inverse series $T(X)$ is obtained by substituting $U(X)$ for Y in the series $1 + \sum_{n>0} Y^n$; moreover, the radius of convergence of the latter is equal to 1 and so $\neq 0$; proposition 6. 1 then follows from proposition 5. 1.

7. Differentiation of a convergent power series

Proposition 7. 1. *Let* $S(X) = \sum_{n \geqslant 0} a_n X^n$ *be a formal power series and let*

$$S'(X) = \sum_{n \geqslant 0} n a_n X^{n-1}$$

be its derived series (cf. § 1, no. 6). *Then the series* S *and* S' *have the same radius of convergence. Moreover, if this radius of convergence* ρ *is* $\neq 0$, *we have, for* $|z| < \rho$,

$$S'(z) = \lim_{h} \frac{S(z+h) - S(z)}{h}, \tag{7. 1}$$

where h *tends to* 0 *without taking the value* 0.

Preliminary remark. If $|z| < \rho$, then $|z+h| < \rho$ for sufficiently small values of h (in fact, for $|h| < \rho - |z|$); thus $S(z+h)$ is defined. On the other hand, it is understood in relation (7. 1) that h tends to 0 through non-zero real values if the field of coefficients is the field **R**, or by non-zero complex values if the field of coefficients is the field **C**. In the case of the field **R**, relation (7. 1) expresses that the function $z \to S(z)$ has *derivative* equal to $S'(z)$; in the case of the complex field **C**, relation (7. 1) shows that we also have the notion of *derivative with respect to the complex variable* z. In both cases, the existence of a derived function $S'(z)$ obviously implies that the function $S(z)$ is continuous for $|z| < \rho$, which can also be proved directly.

Proof of proposition 7. 1. Let $\alpha_n = |a_n|$ and let ρ and ρ' be the radii of convergence of the series S and S′ respectively. If $r < \rho'$, the series $\sum_{n\geq 0} n\alpha_n r^{n-1}$ converges, and so

$$\sum_{n\geq 1} \alpha_n r^n \leqslant r\Big(\sum_{n\geq 0} n\alpha_n r^{n-1}\Big) < +\infty,$$

and, consequently, $r \leqslant \rho$. Conversely, if $r < \rho$ choose an r' such that $r < r' < \rho$; then

$$n\alpha_n r^{n-1} = \frac{1}{r'}(\alpha_n r'^n).n\Big(\frac{r}{r'}\Big)^{n-1};$$

since $r' < \rho$, there exists a finite $M > 0$ such that $\alpha_n r'^n \leqslant M$ for all n, whence

$$n\alpha_n r^{n-1} \leqslant \frac{M}{r} n\Big(\frac{r}{r'}\Big)^{n-1},$$

and, since the series $\sum_{n\geq 1} n\Big(\frac{r}{r'}\Big)^{n-1}$ converges, the series $\sum_{n\geq 1} n\alpha_n r^{n-1}$ also converges; thus $r \leqslant \rho'$. We have then that any number $< \rho'$ is $\leqslant \rho$ and any number $< \rho$ is $\leqslant \rho'$, from which it follows that $\rho = \rho'$.

Relation (7. 1) remains to be proved. Choose a fixed z with $|z| < \rho$ and an r such that $|z| < r < \rho$ and suppose that

$$0 \neq |h| \leqslant r - |z| \tag{7. 2}$$

in what follows.

Then $S(z + h)$ is defined, and we have

$$\frac{S(z+h) - S(z)}{h} - S'(z) = \sum_{n\geq 1} u_n(z, h), \tag{7. 3}$$

where we have put

$$u_n(z, h) = a_n\{(z+h)^{n-1} + z(z+h)^{n-2} + \cdots + z^{n-1} - nz^{n-1}\}.$$

Since $|z|$ and $|z + h|$ are $\leqslant r$, we have $|u_n(z, h)| \leqslant 2n\alpha_n r^{n-1}$; and, since $r < \rho$, we have $\sum_{n\geq 1} n\alpha_n r^{n-1} < +\infty$; thus, given $\varepsilon > 0$, there exists an integer n_0 such that

$$\sum_{n>n_0} 2n\alpha_n r^{n-1} \leqslant \varepsilon/2.$$

With this choice of n_0, the finite sum $\sum_{n\leqslant n_0} u_n(z,h)$ is a polynomial in h which vanishes when $h = 0$; it follows that $\Big|\sum_{n\leqslant n_0} u_n(z,h)\Big| \leqslant \varepsilon/2$ when $|h|$ is smaller than a suitably chosen η. Finally, if h satisfies (7. 2) and $|h| \leqslant \eta$, we deduce from (7. 3) that

$$\left|\frac{S(z+h) - S(z)}{h} - S'(z)\right| \leqslant \Big|\sum_{n\leqslant n_0} u_n(z,h)\Big| + \sum_{n>n_0} 2n\alpha_n r^{n-1} \leqslant \varepsilon.$$

Thus we have proved the relation (7. 1).

Note. It can be shown that the convergence of $\frac{S(z+h)-S(z)}{h}$ towards $S'(z)$ is *uniform* with respect to z for $|z| \leqslant r$ (r being a fixed number strictly less than the radius of convergence ρ).

8. Calculation of the coefficients of a power series

Let $S(x)$ be a formal power series whose radius of convergence $\rho \neq 0$, so that $S(z)$ is the sum of the series $\sum_{n\geqslant 0} a_n z^n$ for $|z| < \rho$. The function $S(z)$ has for derivative the function $S'(z) = \sum_{n\geqslant 0} n a_n z^{n-1}$. We can again apply proposition 7.1 to the series S' to obtain its derived function $S''(z)$, the sum of the power series $\sum_{n\geqslant 0} n(n-1)a_n z^{n-2}$, whose radius of convergence is also ρ. This process can be carried on indefinitely, and by induction we see that the function $S(z)$ is infinitely differentiable for $|z| < \rho$; its derivative of order n is

$$S^{(n)}(z) = n!a_n + T_n(z),$$

where T_n is a series of order $\geqslant 1$, in other words $T_n(0) = 0$. From this, we have

$$a_n = \frac{1}{n!} S^{(n)}(0). \tag{8.1}$$

This fundamental formula shows, in particular, that, if the function $S(z)$ is known in some neighbourhood of 0 (however small), the coefficients a_n of the power series S are completely determined. Consequently, given a function $f(z)$ defined for all sufficiently small $|z|$, *there cannot exist more than one formal power series* $S(X) = \sum_{n\geqslant 0} a_n X^n$ whose radius of convergence is $\neq 0$, and such that $f(z) = \sum_{n\geqslant 0} a_n z^n$ for $|z|$ sufficiently small.

9. Compositional inverse series of a convergent power series.

Refer to § 1, proposition 7.1.

Proposition 9.1. *Let* S *be a power series such that* $S(0) = 0$ *and* $S'(0) \neq 0$, *and let* T *be its inverse series, that is the series such that*

$$T(0) = 0, \qquad S \circ T = I.$$

If the radius of convergence of S *is* $\neq 0$, *then the radius of convergence of* T *is* $\neq 0$. The reader can accept this proposition without proof because a proof (which does not use power series theory) will be given later (chap. IV, § 5, proposition 6.1).

Here, however, a direct proof using power series theory is given to satisfy the reader with an inquisitive mind. It uses the idea of 'majorant series' (cf. chap. VII). Let us keep to the notations of the proof of proposition 7.1 in § 1 and let us consider relations (7. 5) of § 1 which enable us to calculate the unknown coefficients b_n of the required series T(X). Along with the series S(X), we consider a 'majorant' series, that is a series

$$\overline{S}(X) = A_1X - \sum_{n\geqslant 2} A_nX^n$$

with coefficients $A_n > 0$ such that $|a_n| \leqslant A_n$ for all n; moreover we assume that $A_1 = |a_1|$. Applying § 1 proposition 7. 1 to the series $\overline{S}$, gives a series

$$\overline{T}(Y) = \sum_{n\geqslant 1} B_nY^n$$

such that $\overline{S}(\overline{T}(Y)) = Y$; its coefficients B_n are given by the relations

$$A_1B_n - P_n(A_2, \ldots, A_n, B_1, \ldots, B_{n-1}) = 0 \tag{9. 1}$$

which are analogs of (7. 5) of § 1. We obtain from them by induction on n:

$$|b_n| \leqslant B_n. \tag{9. 2}$$

It follows that the radius of convergence of the series T is not less than that of the series $\overline{T}$. We shall prove proposition 9. 1 by showing that the radius of convergence of $\overline{T}$ is > 0.

To this end, we choose the series $\overline{S}$ as follows: let $r > 0$ be a number strictly less than the radius of convergence of the series S (by hypothesis, this radius of convergence is $\neq 0$); the general term of the series $\sum_{n\geqslant 1} |a_n|r^n$ is then bounded above by by a finite number $M > 0$ and, if we put

$$A_1 = |a_1|, \qquad A_n = M/r^n \qquad \text{for } n \geqslant 2, \tag{9. 3}$$

we obtain the coefficients of a majorant series of S; its sum $\overline{S}(x)$ is equal to

$$\overline{S}(x) = A_1x - M\frac{x^2/r^2}{1 - x/r} \qquad \text{for} \qquad |x| < r.$$

We seek, then, a function $\overline{T}(y)$ defined for sufficiently small values of y which is zero for $y = 0$ and which satisfies the equation $\overline{S}(\overline{T}(y)) = y$ identically; $\overline{T}(y)$ must satisfy the quadratic equation

$$(A_1/r + M/r^2)\,\overline{T}^2 - (A_1 + y/r)\overline{T} + y = 0, \tag{9. 4}$$

which has for solution (which vanishes when $y = 0$)

$$\overline{T}(y) = \frac{A_1 + y/r - \sqrt{(A_1)^2 - 2A_1y/r - 4My/r^2 + y^2/r^2}}{2(A_1/r + M/r^2)}.$$

When $|y|$ is sufficiently small, the surd is of the form $A_1\sqrt{1 + u}$, with $|u| < 1$, and so $\overline{T}(y)$ can be expanded as a power series in y, which converges for sufficiently small $|y|$. Thus the radius of convergence of this series is $\neq 0$, as required.

3. Logarithmic and Exponential Functions

1. Exponential function

We have already remarked (§ 2, no. 3) that the formal series $\sum_{n\geqslant 0}\frac{1}{n!}X^n$ has infinite radius of convergence. For z complex, we define

$$e^z = \sum_{n\geqslant 0}\frac{1}{n!}z^n,$$

that is, the sum of an absolutely convergent series. This function has derivative

$$\frac{d}{dz}(e^z) = e^z \tag{1.1}$$

by proposition 7.1 of § 2.

On the other hand, applying proposition 4.2 of § 2 to two series with general terms

$$u_n = \frac{1}{n!}z^n, \qquad v_n = \frac{1}{n!}z'^n,$$

gives

$$w_n = \sum_{0\leqslant p\leqslant n}\frac{1}{p!(n-p)!}z^p z'^{n-p} = \frac{1}{n!}(z+z')^n.$$

Consequently

$$e^{z+z'} = e^z.e^{z'} \tag{1.2}$$

(the fundamental functional property of the exponential function). In particular,

$$e^z.e^{-z} = 1, \quad \text{so} \quad e^z \neq 0 \quad \text{for all } z. \tag{1.3}$$

Putting $z = x + iy$ (with x and y real) gives

$$e^{x+iy} = e^x.e^{iy},$$

so we need only study the two functions e^x and e^{iy}, where x and y are real variables. We have

$$\frac{d}{dx}(e^x) = e^x, \qquad \frac{d}{dy}(e^{iy}) = ie^{iy}. \tag{1.4}$$

2. Real exponential function e^x

We have seen that $e^x \neq 0$: what is more, $e^x = (e^{x/2})^2 > 0$. Moreover, the expansion $e^x = 1 + x + \frac{x^2}{2} + \cdots$ shows that $e^x > 1 + x$ when $x > 0$.

Thus

$$\lim_{x \to +\infty} e^x = +\infty;$$

substituting $-x$ for x leads to

$$\lim_{x \to -\infty} e^x = 0.$$

We deduce that the function e^x of the real variable x increases strictly from 0 to $+\infty$. The transformation $t = e^x$ has therefore a inverse transformation defined for $t > 0$; it is denoted by

$$x = \log t.$$

This function is also strictly monotonic increasing and increases from $-\infty$ to $+\infty$. The functional relation of e^x is written

$$\log(tt') = \log t + \log t', \tag{2.1}$$

and, in particular, $\log 1 = 0$.

On the other hand, the theorem about the derivative of an inverse function gives

$$\frac{d}{dt}(\log t) = 1/t. \tag{2.2}$$

Let us replace t by $1 + u$ $(u > -1)$; $\log(1+u)$ is the primitive of $\frac{1}{1+u}$ which vanishes for $u = 0$; moreover we have the following power series expansion

$$\frac{1}{1+u} = 1 - u + u^2 + \cdots + (-1)^{n-1}u^{n-1} + \cdots$$

whose radius of convergence is equal to 1. From proposition 7.1 of § 2, it follows that the series of the primitive has the same radius of convergence and that its sum has derivative $\frac{1}{1+u}$; whence, for $|u| < 1$,

$$\log(1+u) = u - \frac{u^2}{2} + \cdots + (-1)^{n-1}\frac{u^n}{n} + \cdots \tag{2.3}$$

(in fact this expansion is also correct when $u = 1$).

Now put

$$S(X) = \sum_{n \geq 1} \frac{1}{n!} X^n, \qquad T(Y) = \sum_{n \geq 1} (-1)^{n-1} \frac{Y^n}{n}, \tag{2.4}$$

and examine the composed series $U = S \circ T$. We have from proposition 5.1 of § 2, for $-1 < u < +1$,

$$U(u) = S(T(u));$$

however, $T(u) = \log(1 + u)$, $S(x) = e^x - 1$, so

$$U(u) = e^{\log(1+u)} - 1 = (1 + u) - 1 = u.$$

This shows that the formal series U is merely I because of the uniqueness of the power series expansion of a function (cf. § 2, no. 8). Thus the series S and T are inverse.

3. The imaginary exponential function e^{iy} (y real)

The series expansion of e^{iy} shows that e^{-iy} is the complex conjugate of e^{iy}; thus $e^{iy}.e^{-iy}$ is the square of the modulus of e^{iy}; but this product is equal to 1 by relation (1. 3). Thus

$$|e^{iy}| = 1.$$

We note that, in the Argand plane representation of the complex field $\mathbf{C}$, the point e^{iy} is on the *unit circle*, that is the locus of points whose distance from the origin 0 is equal to 1. The complex numbers u such that $|u| = 1$ form a *group* $\mathbf{U}$ under multiplication and the functional property

$$e^{i(y+y')} = e^{iy}.e^{iy'}$$

expresses the following : *the mapping $y \to e^{iy}$ is a homomorphism of the additive group $\mathbf{R}$ in the multiplicative group $\mathbf{U}$.* This homomorphism will be studied more closely.

Theorem. *The homomorphism $y \to e^{iy}$ maps $\mathbf{R}$ onto $\mathbf{U}$, and its 'kernel' (subgroup of the y such that $e^{iy} = 1$, the neutral element of $\mathbf{U}$) is composed of all the integral multiples of a certain real number > 0. By definition, this number will be denoted by 2π.*

Proof. Let us introduce real and imaginary parts of e^{iy}; we put, by definition,

$$e^{iy} = \cos y + i \sin y,$$

which defines two real functions $\cos y$ and $\sin y$, such that

$$\cos^2 y + \sin^2 y = 1.$$

These functions can be expanded as power series whose radii of convergence are infinite :

$$(3.\ 1)\qquad \begin{cases} \cos y = 1 - \dfrac{1}{2}y^2 + \cdots + \dfrac{(-1)^n}{(2n)!}y^{2n} + \cdots, \\ \sin y = y - \dfrac{1}{3!}y^3 + \cdots + \dfrac{(-1)^n}{(2n+1)!}y^{2n+1} + \cdots. \end{cases}$$

We shall study the way in which these two functions vary. Observe that separating the real and imaginary parts in the second equation (1. 4) gives

$$\frac{d}{dy}(\cos y) = -\sin y, \qquad \frac{d}{dy}(\sin y) = \cos y.$$

When $y = 0$, $\cos y$ is equal to 1; since $\cos y$ is a continuous function, there exists a $y_0 > 0$ such that $\cos y > 0$ for $0 \leqslant y \leqslant y_0$. Hence $\sin y$, whose derivative is $\cos y$, is a strictly increasing function in the interval $[0, y_0]$. Put $\sin y_0 = a > 0$. We shall show that $\cos y$ vanishes for a certain value of y which is > 0. Suppose in fact that $\cos y > 0$ for $y_0 \leqslant y \leqslant y_1$; we have

$$\cos y_1 - \cos y_0 = -\int_{y_0}^{y_1} \sin y \, dy. \tag{3. 2}$$

However, $\sin y \geqslant a$, because $\sin y$ is an increasing function in the interval $[y_0, y_1]$ where its derivative is > 0, thus

$$\int_{y_0}^{y_1} \sin y \, dy \geqslant a(y_1 - y_0).$$

By substituting this in (3. 2) and noting that $\cos y_1 > 0$, we find that

$$y_1 - y_0 < \frac{1}{a} \cos y_0.$$

This proves that $\cos y$ vanishes in the interval $\left[y_0, y_0 + \frac{1}{a}\cos y_0\right]$. Write $\frac{\pi}{2}$ for the smallest value of y which is > 0 and for which $\cos y = 0$ (this is a *definition* of the number π). In the interval $\left[0, \frac{\pi}{2}\right]$, $\cos y$ decreases strictly from 1 to 0, and $\sin y$ increases strictly from 0 to 1; thus the mapping $y \to e^{iy}$ is a bijective mapping of the compact interval $\left[0, \frac{\pi}{2}\right]$ onto the set of points (u, v) of the unit circle whose coordinates u and v are both $\geqslant 0$. By a theorem of topology about continuous, bijective, mappings of a *compact* space, we deduce :

LEMMA. *The mapping $y \to e^{iy}$ is a homeomorphism of $\left[0, \frac{\pi}{2}\right]$ onto the sector of the unit circle $u^2 + v^2 = 1$ in the positive quadrant $u \geqslant 0$, $v \geqslant 0$.*

For $\frac{\pi}{2} \leqslant y \leqslant \pi$, we have $e^{iy} = ie^{i\left(y - \frac{\pi}{2}\right)}$, wnence we easily deduce that e^{iy} takes each complex value of modulus 1 whose abscissa is $\leqslant 0$ and whose ordinate is $\geqslant 0$, and takes each value precisely once. Analogous results can be deduced for the intervals $\left[\pi, \frac{3\pi}{2}\right]$ and $\left[\frac{3\pi}{2}, 2\pi\right]$.

Thus, for $0 \leqslant y < 2\pi$, e^{iy} takes each complex value of modulus 1 precisely once, whereas $e^{2i\pi} = 1$. Therefore the function e^{iy} is periodic of period 2π, and the mapping $y \to e^{iy}$ maps **R** on **U**. This completes the proof of the theorem.

4. Measurement of angles. Argument of a complex number

Let $2\pi\mathbf{Z}$ denote the subgroup of the additive group **R** formed by the integral multiples of the number 2π. The mapping $y \to e^{iy}$ induces an *isomorphism* φ of the *quotient group* $\mathbf{R}/2\pi\mathbf{Z}$ on the *group* **U**. The inverse isomorphism φ^{-1} of **U** on $\mathbf{R}/2\pi\mathbf{Z}$ associates with any complex number u such that $|u| = 1$, a real number which is defined up to addition of an integral multiple of 2π; this class of numbers is called the *argument* of u and is denoted by arg u. By an abuse of notation, arg u will also denote any one of the real numbers whose class modulo 2π is the argument of u; the function arg u is then an example of a many-valued function, that is, it can take many values for a given value of the variable u. This function resolves the problem of ' measure of angles ' (each angle is identified with the corresponding point of **U**) : the ' measure of an angle ' is a real number which is only defined modulo 2π.

We topologize the quotient group $\mathbf{R}/2\pi\mathbf{Z}$ by putting on it the *quotient* topology of the usual topology on the real line **R** : let p be the canonical mapping of **R** on its quotient $\mathbf{R}/2\pi\mathbf{Z}$, a subset A of $\mathbf{R}/2\pi\mathbf{Z}$ is said to be *open* if its inverse image $p^{-1}(A)$, which is a subset of **R** invariant under translation by 2π, is an open set of **R**. It is easily verified that the topological space $\mathbf{R}/2\pi\mathbf{Z}$ is Hausdorff (that is, that two distinct points have disjoint open neighbourhoods). Moreover, it is *compact*; for, if I is the closed interval $[0,2\pi]$, the natural mapping $I \to \mathbf{R}/2\pi\mathbf{Z}$ takes the compact space I onto the Hausdorff space $\mathbf{R}/2\pi\mathbf{Z}$ which is then compact by a classical theorem in topology. The homomorphism $\varphi : \mathbf{R}/2\pi\mathbf{Z} \to \mathbf{U}$ is continuous and is a bijective mapping of the compact space $\mathbf{R}/2\pi\mathbf{Z}$ onto the Hausdorff space **U**; hence φ is a *homeomorphism* of $\mathbf{R}/2\pi\mathbf{Z}$ on **U**.

General definition of argument : for any complex number $t \neq 0$, define the argument of t by the formula

$$\arg t = \arg\left(\frac{t}{|t|}\right).$$

The right hand side is defined already since $t/|t| \in \mathbf{U}$. (Note that the argument of 0 is not defined.) As above, arg t is only defined up to addition of integral multiples of 2π. We thus have

$$t = |t| e^{i \arg t}. \tag{4.1}$$

Application. To solve the equation $t^n = a$ (where $a \neq 0$ is given) : the equation is equivalent to

$$|t| = |a|^{1/n}, \qquad \arg t = \frac{1}{n} \arg a,$$

and has n complex solutions t because one obtains for arg t a real number defined up to addition of an integral multiple of $2\pi/n$.

5. Complex Logarithms

Given a complex number t, we seek all the complex numbers z such that $e^z = t$. Such numbers exist only when $t \neq 0$. In this case, relation (4. 1) shows that the z that we seek are the complex numbers of the form

$$\log |t| + i \arg t. \tag{5. 1}$$

We define

$$\log t = \log |t| + i \arg t, \tag{5. 2}$$

which is a complex number defined only up to addition of an integral multiple of $2\pi i$. From this definition, we have $e^{\log t} = t$. When t is real and > 0, we again have the classical function log t if we allow only the value 0 for arg t.

For any complex numbers t and t' both $\neq 0$ and for any values of log t, log t' and log tt', we have

$$\log (tt') = \log t + \log t' \pmod{2\pi i}. \tag{5. 3}$$

Branches of the logarithm. So far we have not defined log t as a *function* in the proper sense of the word.

Definition. We say that a *continuous* function $f(t)$ of the complex variable t, defined in a *connected open set* D of the plane **C**, not containing the point $t = 0$, is *a branch* of log t if, for all $t \in$ D, we have $e^{f(t)} = t$ (in other words, if $f(t)$ is one of the possible values of log t).

We shall see later (chapter II, § 1, no. 7) what conditions must be satisfied by the open set D for branch of log t to exist in D. We shall now examine how it is possible to obtain all branches of log t if one exists.

Proposition 5. 1 *If there exists a branch* $f(t)$ *of* log t *in the connected open set* D, *then any other branch is of the form* $f(t) + 2k\pi i$ (k *an integer*); *conversely,* $f(t) + 2k\pi i$ *is a branch of* log t *for any integer* k.

Let us suppose the that $f(t)$ and $g(t)$ are two branches of $\log t$. The difference

$$h(t) = \frac{f(t) - g(t)}{2\pi i}$$

is a continuous function in D which takes only integral values; since D is assumed connected, such a function is necessarily *constant*. For, the set of points $t \in D$ such that $h(t)$ is equal to a given integer n is both open and closed. Thus the set is empty or is equal D. The constant must of course be an integer. That $f(t) + 2k\pi i$ is a branch of $\log t$ for any integer k is obvious.

One defines similarly what must be understood by a branch of $\arg t$ in a connected open set D which does not contain the origin. Moreover, any branch of $\arg t$ defines one of $\log t$ and vice-versa.

Example. Let D be the open half-plane $\mathrm{Re}(t) > 0$ (recall that $\mathrm{Re}(t)$ denotes the real part of t). For any t in this half-plane, there is a unique value of $\arg t$ which is $> -\frac{\pi}{2}$ and $< \frac{\pi}{2}$; denote this value by $\mathrm{Arg}\, t$. We shall show that $\mathrm{Arg}\, t$ is a *continuous function* and that consequently

$$\log |t| + i \,\mathrm{Arg}\, t$$

is a branch of $\log t$ in the half plane $\mathrm{Re}(t) > 0$. It will be called the *principal branch* of $\log t$. Since $\mathrm{Arg}\, t = \mathrm{Arg}\,(t/|t|)$ and since the mapping $t \to t/|t|$ is a continuous mapping of the half-plane $\mathrm{Re}(t) > 0$ on the set of u such that $|u| = 1$ and $\mathrm{Re}(u) > 0$, it is sufficient to show that the mapping $y = \mathrm{Arg}\, u$ is continuous. However, this is the inverse mapping of $u = e^{iy}$ as y ranges over the open interval $\left]-\frac{\pi}{2}, +\frac{\pi}{2}\right[$; the function $u = e^{iy}$ is a continuous bijective mapping of the compact interval $\left[-\frac{\pi}{2}, +\frac{\pi}{2}\right]$ on the set of u such that $|u| = 1$ and $\mathrm{Re}(u) \geqslant 0$; this then is a homeomorphism and the inverse mapping is indeed continuous, which completes the proof.

6. Series expansion of the complex logarithm

Proposition 6. 1. *The sum of the power series*

$$T(u) = \sum_{n \geqslant 1} (-1)^{n-1} \frac{u^n}{n},$$

which converges for $|u| < 1$, *is equal to the principal branch of* $\log(1 + u)$.

Note first that if $|u| < 1$, $t = 1 + u$ remains inside an open disc contained in the half plane $\text{Re}(t) > 0$. Again we use the notations of relation (2. 4) and remember that the series S and T are inverse to one another; proposition 5. 1 of § 2 shows that $S(T(u)) = u$ for any complex number u such that $|u| < 1$. In other words, $e^{T(u)} = 1 + u$; and consequently $T(u)$ is a branch of $\log(1 + u)$. To show that this is the principal branch, it is sufficient to verify that it takes the same value as the principal branch for a particular value of u, for instance, that it is zero when $u = 0$, which is obvious from the series expansion of $T(u)$.

PROPOSITION 6. 2. *If $f(t)$ is a branch of* $\log t$ *in a connected open set* D, *the function $f(t)$ has derivative $f'(t)$ with respect to the complex variable t, and*

$$f'(t) = 1/t.$$

In fact, for h complex $\neq 0$ and sufficiently small, we have

$$\frac{f(t + h) - f(t)}{h} = \frac{f(t + h) - f(t)}{e^{f(t+h)} - e^{f(t)}};$$

and, when t tends to 0, this tends to the algebraic inverse of the limit of $\frac{e^{z'} - e^{z}}{z' - z}$ as z' tends to $z = f(t)$; the limit we seek is then the inverse of the value of the derivative of e^z for $z = f(t)$, which is equal to $e^{-f(t)} = 1/t$.

Note. This result checks with the fact that the derivative of the power series $T(u)$ is indeed equal to $\frac{1}{1 + u}$.

Definition. For any pair of complex numbers $t \neq 0$ and α, we put

$$t^{\alpha} = e^{\alpha \log t}.$$

This is a many valued function of t for fixed α. A branch of t^{α} in a connected open set D is defined as above. Any branch of $\log t$ in D defines a branch of t^{α} in D.

Revision. Here the reader is asked to revise, if necessary, the power series expansions of the usual functions, arc tan x, arc sin x, etc. Moreover, for any complex exponent α and for x complex such that $|x| < 1$, we consider

$$(1 + x)^{\alpha} = e^{\alpha \log(1+x)},$$

where $\log(1 + x)$ denotes the principal branch (the function $(1 + x)^{\alpha}$ then takes the value 1 for $x = 0$); the reader should study its power series expansion.

4. Analytic Functions of a Real or Complex Variable

1. Definitions

Definition 1. 1. We say that a function $f(x)$, defined in some neighbourhood of x_0, has a *power series expansion* at the point x_0 if there exists a formal power series $S(X) = \sum_{n \geq 0} a_n X^n$ whose radius of convergence is $\neq 0$ and which satisfies

$$f(x) = \sum_{n \geq 0} a_n (x - x_0)^n \qquad \text{for } |x - x_0| \text{ sufficiently small.}$$

This definition applies equally well to the case when x is a real or a complex variable. The series $S(X)$, if it exists, is *unique* by no. 8 of § 2.

If $f(x)$ has a power series expansion at x_0, then the function f is infinitely differentiable in a neigbourhood of x_0 because the sum of a power series has this property. If the product fg of two functions f and g having power series expansions at x_0 is identically zero in some neighbourhood of x_0, then a least one of the functions f and g is identically zero in a neighbourhood of x_0; in fact, this is an immediate consequence of the fact that the ring of formal series is an integral domain (§ 1, proposition 3. 1). If f has a power series expansion at x, there exists a function g also having a power series expansion at x_0 and having derivative $g' = f$ in some neighbourhood of x_0; such a function is unique up to addition of a constant in some neighbourhood of x_0; to see why this is so, it is sufficient to examine the series of primitives of terms of a power series expansion of the function f.

We shall consider in what follows an *open set* D of the real line $\mathbf{R}$, or the complex plane $\mathbf{C}$. If D is open in $\mathbf{R}$, D is a union of open intervals and, if D is also connected, D is an open interval. We write x for a real or complex variable which varies over the open set D.

Definition 1. 2. A function $f(x)$ with real or complex values defined in the open set D, is said to be *analytic* in D if, for any point $x_0 \in D$, the function $f(x)$ has a power series expansion at the point x_0. In other words, there must exist a number $\rho(x_0) > 0$ and a formal power series $S(X) = \sum_{n \geq 0} a_n X^n$ with radius of convergence $\geq \rho(x_0)$ and such that

$$f(x) = \sum_{n \geq 0} a_n (x - x_0)^n \qquad \text{for} \qquad |x - x_0| < \rho(x_0).$$

The following properties are obvious: any analytic function in D is infinitely differentiable in D and all its derivatives are analytic in D.

The sum and product of two analytic functions in D are analytic in D : that is to say, the analytic functions in D form a ring, and even an algebra. It follows from proposition 6. 1 of § 2 that, if $f(x)$ is analytic in D, then $1/f(x)$ is analytic in the open set D excluding the set of points x_0 such that $f(x_0) = 0$.

Finally, proposition 5. 1 of § 2 gives that, if f is analytic in D and takes its values in D′ and if g is analytic in D′, then the composed function $g \circ f$ is analytic in D.

Let f be an analytic function in a *connected* set D; if f has a primitive g, that is, if there exists a function g in D whose derivative g' is equal to f, then this primitive function is unique up to addition of a constant and it is an analytic function.

Examples of analytic functions. Polynomials in x are analytic functions on the whole of the real line (or in the complex plane). A rational function $P(x)/Q(x)$ is analytic in the complement of the set of points x_0 such that $Q(x_0) = 0$. It will follow from proposition 2. 1 that the function e^x is analytic. The function arc tan x is analytic for all real x since its derivative $\frac{1}{1+x^2}$ is analytic.

2. Criteria of analyticity

PROPOSITION 2. 1. *Let* $S(X) = \sum_{n\geq 0} a_n X^n$ *be a power series whose radius of convergence* ρ *is* $\neq 0$. *Let*

$$S(x) = \sum_{n\geq 0} a_n x^n$$

be its sum for $|x| < \rho$. *Then* $S(x)$ *is an analytic function in the disc* $|x| < \rho$.

This result is by no means trivial. It will be an immediate consequence of what follows, to be precise :

PROPOSITION 2. 2 *With the conditions of proposition* 2. 1, *let* x_0 *be such that* $|x_0| < \rho$. *Then the power series*

$$\sum_{n\geq 0} \frac{1}{n!} S^{(n)}(x_0) X^n \tag{2. 1}$$

has radius of convergence $\geq \rho - |x_0|$ *and*

$$S(x) = \sum_{n\geq 0} \frac{1}{n!} S^{(n)}(x_0)(x-x_0)^n \quad \textit{for} \quad |x-x_0| < \rho - |x_0|. \tag{2. 2}$$

Proof of proposition 2. 2. Put $r_0 = |x_0|$, $\alpha_n = |a_n|$. We have

$$S^{(p)}(x_0) = \sum_{q\geqslant 0} \frac{(p+q)!}{q!} a_{p+q}(x_0)^q,$$
$$|S^{(p)}(x_0)| \leqslant \sum_{q\geqslant 0} \frac{(p+q)!}{q!} \alpha_{p+q}(x_0)^q.$$

For $r_0 \leqslant r < \rho$, we have

$$(2.3)\quad \sum_{p\geqslant 0} \frac{1}{p!} |S^{(p)}(x_0)|\,(r-r_0)^p \leqslant \sum_{p,q} \frac{(p+q)!}{p!\,q!} \alpha_{p+q}(r_0)^q(r-r_0)^p,$$
$$\leqslant \sum_{n\geqslant 0} \alpha_n \Big(\sum_{0\leqslant p\leqslant n} \frac{n!}{p!(n-p)!} (r-r_0)^p (r_0)^{n-p} \Big),$$
$$\leqslant \sum_{n\geqslant 0} \alpha_n r^n < +\infty.$$

Thus the radius of convergence of the series (2. 1) is $\geqslant r - r_0$. Since r can be chosen arbitrarily near to ρ, this radius of convergence is $\geqslant \rho - r_0$.

Now let x be such that $|x - x_0| < \rho - r_0$. The double series

$$\sum_{p,q} \frac{(p+q)!}{p!\,q!} a_{p+q}(x_0)^q(x-x_0)^p$$

is absolutely convergent by (2. 3). Its sum can therefore be calculated by regrouping the terms in an arbitrary manner. We shall calculate this sum in two different ways. A first grouping of terms gives

$$\sum_{n\geqslant 0} a_n \Big(\sum_{0\leqslant p\leqslant n} \frac{n!}{p!\,(n-p)!} (x-x_0)^p(x_0)^{n-p} \Big) = \sum_{n\geqslant 0} a_n x^n = S(x);$$

another grouping gives

$$\sum_{p\geqslant 0} \frac{(x-x_0)^p}{p!} \Big(\sum_{q\geqslant 0} \frac{(p+q)!}{q!} a_{p+q}(x_0)^q \Big) = \sum_{p\geqslant 0} \frac{(x-x_0)^p}{p!} S^{(p)}(x_0).$$

Formula (2. 2) follows from a comparison of these two and this completes the proof.

Note 1. The radius of convergence of series (2. 1) may be strictly larger than $\rho - |x_0|$. Consider, for example, the series

$$S(X) = \sum_{n\geqslant 0} (iX)^n.$$

Then $S(x) = \dfrac{1}{1 - ix}$ for $|x| < 1$. Choose a real number for x_0, so we have

$$\frac{1}{1-ix} = \frac{1}{1-ix_0}\Big(1 - i\frac{x-x_0}{1-ix_0}\Big)^{-1} = \sum_{n\geqslant 0} \frac{i^n}{(1-ix_0)^{n+1}}(x-x_0)^n.$$

This series converges for $|x-x_0|<\sqrt{1+(x_0)^2}$ and $\sqrt{1+(x_0)^2}$ is strictly greater than $1-|x_0|$.

Note 2. Let

$$A(r)=\sum_{n\geqslant 0}|a_n|r^n \qquad \text{for} \qquad r<\rho.$$

From inequality (2. 3), we have

$$\left|\frac{1}{p!}S^{(p)}(x)\right|\leqslant\frac{A(r)}{(r-r_0)^p} \qquad \text{for} \qquad |x|\leqslant r_0<r<\rho. \tag{2. 4}$$

Note 3. If x is a *complex* variable, we shall see in chapter II that any function which is differentiable is analytic and is consequently infinitely differentiable. The situation is completely different in the case of a *real* variable : there exist functions which have a first derivative but no second derivative (one need only consider the primitive of a continuous function which is not differentiable). Moreover, there exist functions which are infinitely differentiable but which are not analytic; here is a simple example : the function $f(x)$, which is equal to zero for $x=0$ and to e^{-1/x^2} for $x\neq 0$, is infinitely differentiable for all x; it vanishes with all its derivatives at $x=0$ so, if it were analytic, it would be identically zero in some neighbourhood of $x=0$, which is not the case.

THEOREM. *In order that an infinitely differentiable function of a real variable x in an open interval D should be analytic in D, it is necessary and sufficient that any point $x_0\in D$ has a neighbourhood V with the following property : there exist numbers M and t, finite and >0, such that*

$$\left|\frac{1}{p!}f^{(p)}(x)\right|\leqslant M.t^p \qquad \textit{for any } x\in V \textit{ and any integer } p\geqslant 0. \tag{2. 5}$$

Indication of proof. The condition is shown to be necessary by using inequality (2. 4). It is shown to be sufficient by writing a finite Taylor expansion of the function $f(x)$ and using (2. 5) to find an upper bound for the Lagrange remainder.

3. PRINCIPLE OF ANALYTIC CONTINUATION

THEOREM. *Let f be an analytic function in a connected open set D and let $x_0\in D$. The following conditions are equivalent :*

a) $f^{(n)}(x_0)=0$ *for all integers* $n\geqslant 0$;
b) *f is identically zero in a neighbourhood of x_0;*
c) *f is identically zero in D.*

Proof. It is obvious that *c*) implies *a*). We shall show that *a*) implies *b*) and *b*) implies *c*). Suppose *a*) is satisfied. We have then $f^{(n)}(x_0) = 0$ for all $n \geqslant 0$ with the convention that $f^{(0)} = f$. But $f(x)$ has a power series expansion in powers of $(x - x_0)$ in a neigbourhood of x_0 and the coefficients $\frac{1}{n!} f^n(x_0)$ are zero; thus $f(x)$ is identically zero in a neighbourhood of x_0 which proves *b*).
Suppose conditions *b*) is satisfied. To show that f is zero at all points of D, it is sufficient to show that the set D′ of points $x \in$ D *in a neighbourhood of which f is identically zero* is both open and closed (D′ is not empty because of *b*), thus, since D is connected, D′ will be equal to D). It follows from the definition of D′ that it is open. It remains to be proved that, if $x_0 \in$ D is in the closure of D′, then $x_0 \in$ D′. However, $f^{(n)}(x) = 0$ for each $n \geqslant 0$ at points arbitrarily close to x_0 (in fact, at the points of D′); thus $f^{(n)}(x_0) = 0$ because of the continuity of $f^{(n)}$; this holding for all $n \geqslant 0$ implies as above that $f(x)$ is identically zero in a neighourhood of x_0. Thus $x_0 \in$ D′, which completes the proof.

Corollary 1. *The ring of analytic functions in a connected open set* D *is an integral domain.*

For, if the product fg of two analytic functions in D is identically zero and if $x_0 \in$ D, then one of the functions f, g is identically zero in a neighbourhood of x_0 because the ring of formal power series is an integral domain. But, if f is identically zero in some neighbourhood of x_0, then f is zero in the whole of D by the above theorem.

Corollary 2. (Principle of analytic continuation) *If two analytic functions f and g in a connected open set* D *coincide in a neighbourhood of a point of* D, *then they are identical in* D.

The *problem of analytic continuation* is the following : given an analytic function h in a connected open set D′ and given a connected open set D containing D′, we ask if there exists an analytic function f in D which extends h. Corollary 2 shows that such a function f is unique if it exists.

4. Zeros of an analytic function

Let $f(x)$ be an analytic function in a neighbourhood of x_0 and let

$$f(x) = \sum_{n \geqslant 0} a_n (x - x_0)^n$$

be its power series expansion for sufficiently small $|x - x_0|$. Suppose that $f(x_0) = 0$ and that $f(x)$ *is not identically zero* in a neighbourhood of x_0.

Let k be the smallest integer such that $a_k \neq 0$. The series

$$\sum_{n \geqslant k} a_n (x - x_0)^{n-k}$$

converges for sufficiently small $|x - x_0|$ and its sum $g(x)$ is an analytic function such that $g(x_0) \neq 0$ in some neighbourhood of x. Thus, for x near enough to x_0, we have

$$(4.1) \qquad f(x) = (x - x_0)^k g(x), \qquad g(x_0) \neq 0.$$

The integer $k > 0$ thus defined is called the *order of multiplicity* of the zero x_0 for the function f. It is characterized by relation (4. 1), where $g(x)$ is analytic in a neighbourhood of x_0. The order of multiplicity k is also characterized by the condition

$$f^{(n)}(x_0) = 0 \quad \text{for} \quad 0 \leqslant n < k, \quad f^{(k)}(x_0) \neq 0.$$

If $k = 1$, we call x_0 a *simple* zero. If $k \geqslant 2$, we call x_0 a *multiple* zero.

Relation (4. 1) and continuity of $g(x)$ imply

$$f(x) \neq 0 \qquad \text{for} \qquad 0 < |x - x_0| < \varepsilon \qquad (\varepsilon > 0 \text{ sufficiently small}).$$

In other words the point x_0 has a neighbourhood in which it is the *unique* zero of the function $f(x)$.

PROPOSITION 4. 1. *If f is an analytic function in a connected open set* D *and if f is not identically zero, then the set of zeros of f is a discrete set* (in other words, all the points of this set are *isolated*).

For, corollary 2 of no. 3 gives that f is not identically zero in a neigbourhood of any point of D, so one can apply the above reasoning to each zero of f.

In particular, any *compact* subset of D contains only a *finite* number of zeros of the function g.

5. MEROMORPHIC FUNCTIONS

Let f and g be two analytic functions in a connected open set D, and suppose that g is not identically zero. The function $f(x)/g(x)$ is defined and analytic in a neighbourhood of every point x_0 of D such that $g(x_0) \neq 0$, that is to say, in the whole of D except perhaps in certain isolated points.

Let us see how $f(x)/g(x)$ behaves in a neighbourhood of a point x_0 which is a zero of $g(x)$; if $f(x)$ is not identically zero, we have

$$f(x) = (x - x_0)^k f_1(x), \qquad g(x) = (x - x_0)^{k'} g_1(x),$$

where k and k' are integers with $k \geqslant 0$ and $k' > 0$, f_1 and g_1 are analytic in some neighbourhood of x_0 with $f_1(x_0) \neq 0$ and $g_1(x_0) \neq 0$; hence, for $x \neq x_0$ but near to x_0,

$$\frac{f(x)}{g(x)} = (x - x_0)^{k-k'} \frac{f_1(x)}{g_1(x)}.$$

The function $h_1(x) = f_1(x)/g_1(x)$ is analytic in a neighbourhood of x_0 and we have that $h_1(x_0) \neq 0$. Two cases arise :

1° $k \geqslant k'$; then the function

$$(x - x_0)^{k-k'} h_1(x)$$

is analytic in some neighbourhood of x_0 and coincides with $f(x)/g(x)$ for $x \neq x_0$. Hence the extension of f/g to the point x_0 is analytic in a neighbourhood of x_0 and admits x_0 as a zero if $k > k'$.

2° $k < k'$: then

$$\frac{f(x)}{g(x)} = \frac{1}{(x - x_0)^{k'-k}} h_1(x), \qquad h_1(x_0) \neq 0.$$

We say in this case that x_0 is a *pole* of the function f/g; the integer $k' - k$ is called the *order of multiplicity* of the pole. As x tends to x_0, $\left|\frac{f(x)}{g(x)}\right|$ tends to $+\infty$. We can agree to extend the function f/g by giving it the value " infinity " at x_0. We shall return later to the introduction of this unique number infinity, denoted ∞.
If $f(x)$ analytic and has x_0 as a zero of order $k > 0$, then x_0 is clearly a pole of order k of $1/f(x)$.

Definition. A *meromorphic* function in an open set D is defined to be a function $f(x)$ which is defined and analytic an the open set D′ obtained from D by taking out a set of isolated points each of which is a *pole* of $f(x)$.

In a neighbourhood of each point of D (without exception), f can be expressed as a quotient $h(x)/g(x)$ of two analytic functions, the denominator being not identically zero. The sum and product of two meromorphic functions are defined in the obvious way : the meromorphic functions in D form a ring and even an algebra. In fact they form a field because, if $f(x)$ is not identically zero in D, it is not identically zero in any neighbourhood of any point of D by the theorem of no. 3; so $1/f(x)$ is then analytic, or has at most a pole at each point of D and is consequently meromorphic in D.

Proposition 5. 1. *The derivative f' of a meromorphic function f in D is meromorphic in D; the functions f and f' have the same poles; if x_0 is a pole of order k of f, then it is a pole of order $k+1$ of f'.*

For, f' is defined and analytic at each point of D which is not a pole of f. It remains to be proved that, if x_0 is a pole of f, x_0 is also a pole of f'. Moreover, for x near x_0,

$$f(x) = \frac{1}{(x - x_0)^k} g(x),$$

$g(x)$ being analytic with $g(x_0) \neq 0$, $k > 0$. Hence, for $x \neq x_0$,

$$f'(x) = \frac{1}{(x - x_0)^{k+1}} [(x - x_0)g'(x) - kg(x)] = \frac{1}{(x - x_0)^{k+1}} g_1(x),$$

and as $g_1(x_0) \neq 0$, x_0 is a pole of f' of order $k + 1$.

Exercises

1. Let K be a commutative field, X an indeterminate and E = K[[X]] the algebra of formal power series with coefficients in K. For S, T in E, define

$$d(S, T) = \begin{cases} 0 & \text{if } S = T, \\ e^{-k} & \text{if } S \neq T, \end{cases} \quad \text{and} \quad \omega(S - T) = k.$$

a) Show that d defines a distance function in the set E.

b) Show that the mappings $(S, T) \to S + T$ and $(S, T) \to ST$ of $E \times E$ into E are continuous with respect to the metric topology defined by d.

c) Show that the algebra K[X] of polynomials is everywhere dense in E when considered as a subset of E.

d) Show that the metric space E is complete. (If (S_n) is a Cauchy sequence in E, note that for any integer $m > 0$, the first m terms of S_n do not depend on n for sufficiently large n.)

e) Is the mapping $S \to S'$ (the derivative of S) continuous?

2. Let p, q be integers $\geqslant 1$. Let $S_1(X)$ be the formal series

$$1 + X + X^2 + \cdots + X^n + \cdots,$$

and put

$$S_p(X) = (S_1(X))^p.$$

a) Show, by induction on n, that

$$(1) \quad 1 + p + \frac{p(p+1)}{2!} + \cdots + \frac{p(p+1)\dots(p+n-1)}{n!} = \frac{(p+1)\dots(p+n)}{n!},$$

and deduce (by induction on p), the expansion

$$S_p(X) = \sum_{n\geqslant 0} \binom{p+n-1}{n} X^n, \tag{2}$$

where $\binom{k}{h}$ denotes the binomial coefficient $\dfrac{k!}{h!(k-h)!}$

b) Use $S_p(X).S_q(X) = S_{p+q}(X)$ to show that

$$\sum_{0\leqslant l\leqslant n} \binom{p+l+1}{l}\binom{q+n+l-1}{n-l} = \binom{p+q+n+1}{n} \tag{3}$$

(which is a generalisation of (1), the case when $q = 1$).

3. Find the precise form of the polynomials P_n in the proof of proposition 7. 1, § 1, for $n \leqslant 5$ and calculate the terms of degree $\leqslant 5$ of the formal (compositional) inverse series of

$$S(X) = X - \frac{1}{3}X^3 + \frac{1}{5}X^5 + \cdots + (-1)^p \frac{1}{2p+1} X^{2p+1} + \cdots.$$

4. Find the radii of convergence of the following series :

a) $$\sum_{n\geqslant 0} q^{n^2} z^n \qquad (|q| < 1),$$

b) $$\sum_{n\geqslant 0} n^p z^n \qquad (p \text{ integer} > 0),$$

c) $$\sum_{n\geqslant 0} a_n z^n, \text{ with } a_{2n+1} = a^{2n+1}, \quad a_{2n} = b^{2n} \text{ for } n > 0,$$

where a and b are real and $0 < a,\ b < 1$.

5. Given two formal power series

$$S(X) = \sum_{n\geqslant 0} a_n X^n \quad \text{and} \quad T(X) = \sum_{n\geqslant 0} b_n X^n \quad (b_n \neq 0),$$

let

$$U(X) = \sum_{n\geqslant 0} (a_n)^p X^n, \quad V(X) = \sum_{n\geqslant 0} a_n b_n X^n, \quad W(X) = \sum_{n\geqslant 0} (a_n/b_n) X$$

(where p is an integer). Prove the following relations :

$$\rho(U) = (\rho(S))^p, \qquad \rho(V) \geqslant \rho(S).\rho(T),$$

and, if $\rho(T) \neq 0$,

$$\rho(W) \leqslant \rho(S)/\rho(T).$$

6. Let a, b and c be elements of $\mathbf{C}$, c not an integer $\leqslant 0$. What is the radius of convergence of the series

$$S(X) = 1 + \frac{ab}{c}X + \frac{a(a+1).(b+1)}{2!c(c+1)}X^2 + \cdots$$
$$+ \frac{a(a+1)\ldots(a+n-1).b(b+1)\ldots(b+n-1)}{n!c(c+1)\ldots(c+n-1)}X^n + \cdots$$

Show that its sum $S(z)$, for $|z| < \rho(S)$, satisfies the differential equation

$$z(1-z)S'' + (c-(a+b+1)z)S' - abS = 0.$$

7. Let $S(X) = \sum_{n\geqslant 0} a_n X^n$ be a formal power series such that $\rho(S) = 1$. Put

$$s_n = a_0 + \cdots + a_n, \quad t_n = \frac{1}{n+1}(s_0 + s_1 + \cdots + s_n) \quad \text{for} \quad n \geqslant 0,$$

and put

$$U(X) = \sum_{n\geqslant 0} s_n X^n, \qquad V(X) = \sum_{n\geqslant 0} t_n X^n.$$

Show that : (i) $\rho(U) = \rho(V) = 1$, (ii) for all $|z| < 1$,

$$\frac{1}{1-z}\left(\sum_{n\geqslant 0} a_n z^n\right) = \sum_{n\geqslant 0} s_n z^n.$$

8. Let $S(X) = \sum_{n\geqslant 0} a_n X^n$ be a formal power series whose coefficients are defined by the following recurrence relations :

$$a_0 = 0, \ a_1 = 1, \ a_n = \alpha a_{n-1} + \beta a_{n-2} \quad \text{for} \quad n \geqslant 2,$$

where α, β are given real numbers.

a) Show that, for $n \geqslant 1$, we have $|a_n| \leqslant (2c)^{n-1}$ where $c = \max(|\alpha|, |\beta|, 1/2)$ and deduce that the radius of convergence $\rho(S) \neq 0$.

b) Show that

$$(1 - \alpha z - \beta z^2)S(z) = z, \quad \text{for} \quad |z| < \rho(S),$$

and deduce that, for $|z| < \rho(S)$,

(1) $$S(z) = \frac{z}{1 - \alpha z - \beta z^2}.$$

c) Let z_1, z_2 be the two roots of $\beta X^2 + \alpha X - 1 = 0$. By decomposing the right hand side of (1) into partial fractions, find an expression for the a_n in terms of z_1 and z_2 and deduce that

$$\rho(S) = \min(|z_1|, |z_2|).$$

(Note that, if $S(X) = S_1(X) . S_2(X)$, then $\rho(S) \geqslant \min(\rho(S_1), \rho(S_2))$.)

9. Show that, if x, y are real and n is an integer $\geqslant 0$, then

$$\sum_{0 \leqslant p \leqslant n} \sin(px + y) = \sin\left(\frac{n}{2}x + y\right) \sin\frac{n+1}{2}x / \sin\frac{x}{2}.$$

$$\sum_{0 \leqslant p \leqslant n} \cos(px + y) = \cos\left(\frac{n}{2}x + y\right) \sin\frac{n+1}{2}x / \sin\frac{x}{2},$$

(Use $\cos(px + y) + i\sin(px + y) = e^{i(px+y)} = e^{iy}(e^{ix})^p$.)

10. Prove the following inequalities for $z \in \mathbf{C}$:

$$|e^z - 1| \leqslant e^{|z|} - 1 \leqslant |z| e^{|z|}.$$

11. Show that, for any integer $n \geqslant 1$ and any complex number z,

$$\left(1 + \frac{z}{n}\right)^n = 1 + z + \sum_{2 \leqslant p \leqslant n} \left(1 - \frac{1}{n}\right) \cdots \left(1 - \frac{p-1}{n}\right) \frac{z^p}{p!},$$

and deduce that

$$e^z = \lim_{n \to \infty} \left(1 + \frac{z}{n}\right)^n.$$

12. Show that the function of a complex variable z defined by

$$\cos z = \frac{e^{iz} + e^{-iz}}{2} \quad \left(\text{resp. } \sin z = \frac{e^{iz} - e^{-iz}}{2i}\right)$$

is the analytic extension to the whole plane $\mathbf{C}$ of the function $\cos x$ (resp. $\sin x$) defined in § 3, no. 3, Prove that, for any $z, z' \in \mathbf{C}$,

$$\cos(z + z') = \cos z \cos z' - \sin z \sin z',$$
$$\sin(z + z') = \sin z \cos z' + \cos z \sin z';$$

$$\cos^2 z + \sin^2 z = 1.$$

13. Prove the relations

$$\frac{2}{\pi}x \leqslant \sin x \leqslant x \quad \text{for } x \text{ real and } 0 \leqslant x \leqslant \pi/2.$$

14. Let $z = x + iy$ with x, y real.

(i) Show that

$$|\sin(x+iy)|^2 = \sin^2 x + \sinh^2 y,$$
$$|\cos(x+iy)|^2 = \cos^2 x + \sinh^2 y;$$

(ii) determine the zeros of the functions $\sin az$, $\cos az$ (where a is a real number $\neq 0$);

(iii) Show that, if $-\pi < a < \pi$ and n is a positive integer,

$$\left|\frac{\sin az}{\sin \pi z}\right| \leqslant \frac{\cosh ay}{\cosh \pi y}, \quad \text{for} \quad z = n + \frac{1}{2} + iy,$$

and

$$\left|\frac{\sin az}{\sin \pi z}\right| \leqslant \frac{\cosh a\left(n+\frac{1}{2}\right)}{\sinh \pi\left(n+\frac{1}{2}\right)}, \quad \text{for} \quad z = x + i\left(n+\frac{1}{2}\right).$$

(N. B. By definition, $\cosh z = \cos(iz)$, $\sinh z = -i\sin(iz)$.)

15. Let I be an interval of the real line **R**. Show that, if $f(x)$ is an analytic function (of a real variable but with complex values) in I, it can be extended to an analytic function in a connected open set D of the complex plane containing I.

16. (i) Let (α_n), (β_n) be two sequences of numbers with the following properties :

a) there is a constant $M > 0$ such that

$$|\alpha_1 + \alpha_2 + \cdots + \alpha_n| \leqslant M \quad \text{for all} \quad n \geqslant 1,$$

b) the β_n are real $\geqslant 0$ and $\beta_1 \geqslant \beta_2 \geqslant \cdots \geqslant \beta_n \geqslant \cdots$.

Show that, for all $n \geqslant 1$,

$$|\alpha_1\beta_1 + \alpha_2\beta_2 + \cdots + \alpha_n\beta_n| \leqslant M\beta_1.$$

(Introduce $s_n = \alpha_1 + \cdots + \alpha_n$ and write

$$\alpha_1\beta_1 + \cdots + \alpha_n\beta_n = (\beta_1 - \beta_2)s_1 + \cdots + (\beta_{n-1} - \beta_n)s_{n-1} + \beta_n s_n.)$$

(ii) Let $S(X) = \sum_{n\geqslant 0} a_n X^n$ be a formal power series with complex coefficients such that $\rho(S) = 1$, and that $\sum_{n\geqslant 0} a_n$ is convergent. Use (i) to show that the series $\sum_{n\geqslant 0} a_n x^n$ is uniformly convergent in the closed interval $[0, 1]$ of **R**, and deduce that

$$\lim_{\substack{x\to 1\\ 0<x<1}} \sum_{n\geqslant 0} a_n x^n = \sum_{n\geqslant 0} a_n.$$

(iii) Let $S(X) = \sum_{n \geq 1} X^n/n^2$ now and let D be the intersection of the open disc $|z| < 1$ and of the open disc $|z - 1| < 1$. Show that there exists a constant a such that

$$S(z) + S(1 - z) = a - \log z \log (1 - z) \quad \text{for} \quad z \in D,$$

where log denotes the principal branch of the complex logarithm in the half-plane $\operatorname{Re}(z) > 0$ (which contains D).

(Note that, if $z \in D$, then $\log (1 - z) = - T(z)$ with

$$T(X) = X . S'(X),$$

because of proposition 6. 1 of § 3, and that proposition 6. 2 of § 3 gives

$$\frac{d}{dz} (\log z \log (1 - z)) = \frac{\log (1 - z)}{z} - \frac{\log z}{1 - z} \quad \text{for} \quad z \in D.)$$

Finally, use (ii) to show that

$$a = \sum_{n \geq 1} 1/n^2,$$

$$a - (\log 2)^2 = \sum_{n \geq 1} 1/n^2 2^{n-1}.$$

(Cf. chapter v, § 2, no. 2, the application of proposition 2. 1.)

CHAPTER II

Holomorphic Functions, Cauchy's Integral

1. Curvilinear Integrals

1. GENERAL THEORY

We shall revise some of the elementary ideas in the theory of curvilinear integrals in the plane $\mathbf{R}^2$. Let x and y denote the coordinates in $\mathbf{R}^2$.

A *differentiable path is a mapping*

$$t \to \gamma(t) \tag{1.1}$$

of the segment $[a, b]$ into the plane $\mathbf{R}^2$, such that the coordinates $x(t)$ and $y(t)$ of the point $\gamma(t)$ are continuously differentiable functions. We shall always suppose that $a < b$. The *initial point* of γ is $\gamma(a)$ and its *end point* is $\gamma(b)$. If D is an open set of the plane, we say that γ is a differentiable path of the open set D if the function γ takes its values in D.

A *differential form* in an open set D is an expression

$$\omega = P\,dx + Q\,dy$$

whose coefficients P and Q are (real- or complex-valued) continuous functions in D.

If γ is a differentiable path of D and ω a differential form in D, we define the integral $\int_\gamma \omega$ by the formula

$$\int_\gamma \omega = \int_a^b \gamma^*(\omega),$$

where $\gamma^*(\omega)$ denotes the differential form $f(t)\,dt$ defined by

$$f(t) = P(x(t), y(t))\,x'(t) + Q(x(t), y(t))\,y'(t);$$

in other words, $\gamma^*(\omega)$ is the differential form deduced from ω by the change of variables $x = x(t)$, $y = y(t)$. Thus,

$$\int_\gamma \omega = \int_a^b f(t)\, dt.$$

Consider now a continuously differentiable function $t = t(u)$ for $a_1 \leqslant u \leqslant b_1$ (with $a_1 < b_1$), whose derivative $t'(u)$ is always > 0 and which is such that $t(a_1) = a$, $t(b_1) = b$. The composed mapping of $u \to t(u)$ and the mapping (1. 1) is

$$(1.\ 2) \qquad u \to \gamma(t(u)).$$

It defines a differentiable path γ_1. We say that γ_1 is deduced from γ by *change of parameter*. The differential form $f_1(u)\, du$ deduced from ω by the mapping (1. 2) is equal to

$$f(t(u))t'(u)\, du,$$

by virtue of the formula giving the derivative of a composed function. The formula for change of variable in an ordinary integral thus gives the equation

$$\int_\gamma \omega = \int_{\gamma_1} \omega.$$

In other words, the curvilinear integral $\int_\gamma \omega$ does not change its value if the differentiable path γ is replaced by another which is deduced from γ by change of parameter. We can, then, denote paths deduced from one another by change of parameter by the same symbol.

Take now a continuously differentiable function $t = t(u)$ defined for $a_1 \leqslant u \leqslant b_1$, but such that $t'(u) < 0$, $t(a_1) = b$, $t(b_1) = a$ (the description of the segment is reversed). We then see that $\int_{\gamma_1} \omega = -\int_\gamma \omega$. We say therefore that we have made a change of parameter in γ which *changes the orientation* of γ; the effect of this is to multiply $\int_\gamma \omega$ by -1.

Subdivide the interval $[a, b]$ described by the parameter t into a finite number of sub-intervals

$$[a, t_1], \quad [t_1, t_2], \quad \ldots, \quad [t_{n-1}, t_n], \quad [t_n, b],$$

where $a < t_1 < t_2 < \ldots < t_{n-1} < t_n < b$. Let γ_i be the restriction of the mapping γ to the i-th of these intervals; it is clear that

$$\int_\gamma \omega = \sum_{i=1}^{n} \left(\int_{\gamma_i} \omega \right).$$

This result leads to a generalization of the idea of a differentiable path. A *piecewise differentiable path is defined to be a continuous mapping*

$$\gamma : [a, b] \to \mathbf{R}^2,$$

such that there exists a subdivision of the interval $[a, b]$ into a finite number of sub-intervals as above, with the property that the restriction of γ to each sub-interval is continuously differentiable. We define

$$\int_\gamma \omega = \sum_{i=1}^{n+1} \left(\int_{\gamma_i} \omega \right).$$

The sum on the right hand side is independent of the decomposition. The initial point of γ_1 is called the initial point of γ and the final point of γ_{n+1} is called the final point of γ. We say that a path is *closed* if its initial and final points coincide.

A closed path γ can also be defined by taking, instead of a real parameter t varying from a to b, a parameter θ which describes the unit circle.

Example. Consider, in the plane $\mathbf{R}^2$, the perimeter (or 'boundary') of a rectangle A whose sides are parallel to the coordinate axes. The rectangle is the set of points (x, y) satisfying

$$a_1 \leqslant x \leqslant a_2, \qquad b_1 \leqslant y \leqslant b_2.$$

Its boundary consists of the four line segments

$$\begin{aligned} x &= a_2, & b_1 &\leqslant y \leqslant b_2, \\ y &= b_2, & a_1 &\leqslant x \leqslant a_2, \\ x &= a_1, & b_1 &\leqslant y \leqslant b_2, \\ y &= b_1, & a_1 &\leqslant x \leqslant a_2. \end{aligned}$$

For this boundary to define a piecewise differentiable closed path γ, it is necessary to stipulate the sense of description chosen. We agree always take the following sense of description :

y increases from b_1 to b_2, along the side $x = a_2$,
x decreases from a_2 to a_1, along the side $y = b_2$,
y decreases from b_2 to b_1, along the side $x = a_1$,
x increases from a_1 to a_2, along the side $y = b_1$.

Thus the integral $\int_\gamma \omega$ is well-defined; it does not depend on the choice of the initial point of γ because it is always equal to the sum of integrals along the four sides, each described in the sense indicated.

2 Primitive of a differential form

Lemma. *Let D be a connected open set of the plane. Any two points $a \in D$ and $b \in D$ are the initial and final points, respectively, of some piecewise differentiable path in D.* (Briefly this says that a and b can be joined by a piecewise differentiable path).

Proof. Each point $c \in D$ is the centre of a disc contained in D and can be joined to each point of this disc by a piecewise differentiable path contained in D, for instance, a radius. Suppose that $a \in D$ is a given point; if c can be joined to a, then any point sufficiently near to c can also be joined to a because of the previous remark; thus the set E of points of D which can be joined to a is *open*. On the other hand, E is closed in D; because, if $c \in D$ is in the closure of E, c can be joined to some point of E because of previous remarks, so c can be joined to a. By hypothesis, D is connected; the subset E of D is non-empty (as $a \in E$) and is both open and closed, so it must be the whole of D. This completes the proof.

Let D again be a connected open set in the plane and let γ be a piecewise differentiable path contained in D with initial point a and final point b. Let F be a continuously differentiable function in D and consider the differential form $\omega = dF$; then we have the obvious relation

$$\int_\gamma dF = F(b) - F(a). \tag{2.1}$$

It follows from this and the lemma that, *if the differential* dF *is identically zero in* D, *the function* F *is constant in* D.

Given a differential form ω in a connected open set D, we investigate whether or not there is a continuously differentiable function $F(x, y)$ in D such that $dF = \omega$. If $\omega = P\,dx + Q\,dy$, the relation $dF = \omega$ is equivalent to

$$\frac{\partial F}{\partial x} = P, \qquad \frac{\partial F}{\partial y} = Q. \tag{2.2}$$

Such a function F, if it exists, is called *a primitive* of the form ω. In this case, any other primitive G is obtained by adding a constant to F since $d(F - G) = 0$.

Proposition 2.1. *A necessary and sufficient condition that a differential form* ω *has a primitive in* D *is that* $\int_\gamma \omega = 0$ *for any piecewise differentiable closed path* γ *contained in* D.

Proof. 1. The condition is necessary because, if $\omega = dF$, relation (2.1) shows that $\int_\gamma \omega = 0$ whenever the initial and final points of γ coincide.

2. The condition is sufficient. For, choose a point $(x_0, y_0) \in D$; any point $(x, y) \in D$ can be joined to (x_0, y_0) by a piecewise continuously differentiable path γ contained in D (by the lemma); the integral $\int_\gamma \omega$ does not depend on the choice of γ because the integral of ω round any closed path is zero by hypothesis. Let $F(x,y)$ be the common value of the integrals $\int_\gamma \omega$ along paths γ in D with initial point (x_0, y_0) and final point (x, y). We shall show that the function F so defined in D satisfies relations (2. 2).

Give x a small increment h; the difference

$$F(x + h, y) - F(x, y)$$

is equal to the integral $\int \omega$ along any path contained in D starting at (x, y) and ending at $(x + h, y)$. In particular, let us integrate along the line segment parallel to the x-axis (which is possible if $|h|$ is small enough):

$$F(x + h, y) - F(x, y) = \int_x^{x+h} P(\xi, y)\, d\xi,$$

and consequently, if $h \neq 0$,

$$\frac{F(x + h, y) - F(x, y)}{h} = \frac{1}{h} \int_x^{x+h} P(\xi, y)\, d\xi.$$

As h tends to 0, the right hand side tends to $P(x, y)$ because of the continuity of the function P. Hence we indeed have

$$\frac{\partial F}{\partial x} = P(x, y).$$

We could prove $\frac{\partial F}{\partial y} = Q(x, y)$ similarly. This completes the proof of proposition 2. 1.

Consider in particular the rectangles contained in D whose sides are parallel to the axes (we mean that the rectangle must be entirely contained in D, both its interior and its frontier). If γ is the boundary of such a rectangle, we must have $\int_\gamma \omega = 0$ for the differential form ω to have a primitive in D. This necessary condition is not always sufficient as we shall see later. Nevertheless, it is sufficient when D is 'simply connected' (cf. no. 7). For the moment we shall confine ourselves to proving following :

Proposition 2. 2. *Let* D *be an open disc. If* $\int_\gamma \omega = 0$ *whenever* γ *is the boundary of a rectangle contained in* D *with sides parallel to the axes, then* ω *has a primitive in* D.

Proof. Let (x_0, y_0) be the centre of the disc D and let (x, y) be a general point of D. There are two paths γ_1 and γ_2 starting at (x_0, y_0) and ending at (x, y), each of which is composed of two sides of the rectangle (with sides parallel to the axes) whose opposite corners are (x_0, y_0) and (x, y) [see figure 1]. Thus this rectangle is contained in D and $\int_{\gamma_1} \omega = \int_{\gamma_2} \omega$. Let

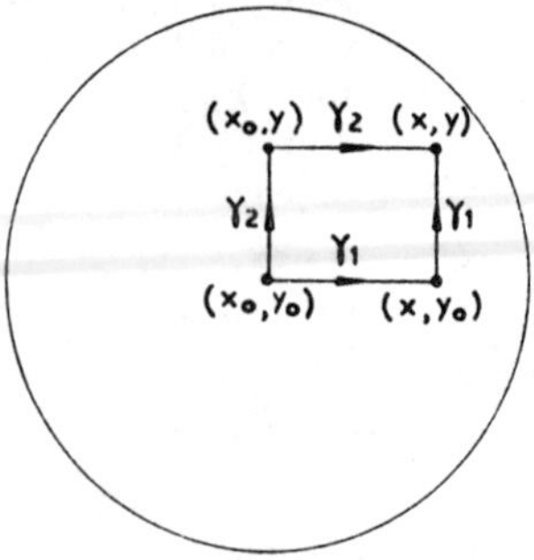

Fig. 1.

$F(x, y)$ be the common value of these two integrals; then we can show as above, that $\frac{\partial F}{\partial x} = P$, $\frac{\partial F}{\partial y} = Q$, which proves the proposition.

3. The Green-Riemann formula

This formula, in some sense, generalizes relation (2. 1) : instead of relating the value of an ordinary integral to values of a function, it relates the value of a double integral to that of a curvilinear one. Let A be a rectangle with sides parallel to the axes, let γ be its boundary and let $P(x, y)$ and $Q(x, y)$ be continuous functions defined in a neighbourhood D of A, the functions having continuous partial derivatives $\frac{\partial P}{\partial y}$ and $\frac{\partial Q}{\partial x}$.

The Green-Riemann formula can then be written

$$\int_\gamma P\,dx + Q\,dy = \iint_A \left(\frac{\partial Q}{\partial x} - \frac{\partial P}{\partial y}\right) dx\,dy. \tag{3. 1}$$

Proof. We shall prove for instance that

$$\int_\gamma Q\,dy = \iint_A \frac{\partial Q}{\partial x}\,dx\,dy.$$

We know that the double integral of the continuous function $\frac{\partial Q}{\partial x}$ can be calculated as follows :

$$\iint_A \frac{\partial Q}{\partial x}\,dx\,dy = \int_{b_1}^{b_2} dy \left(\int_{a_1}^{a_2} \frac{\partial Q}{\partial x}\,dx\right).$$

However, $\int_{a_1}^{a_2} \frac{\partial Q}{\partial x} dx = Q(a_2, y) - Q(a_1, y)$; integrating this with respect to y gives

$$\int_{b_1}^{b_2} Q(a_2, y)\, dy - \int_{b_1}^{b_2} Q(a_1, y)\, dy$$

which is precisely equal to $\int_\gamma Q\, dy$.

This completes the proof.

The Green-Riemann formula is valid for more general domains than rectangles, but we shall leave this question aside for the moment.

PROPOSITION 3. 1. *Let $\omega = P\,dx + Q\,dy$ be a differential form in a connected open set D, and suppose that the partial derivatives $\frac{\partial P}{\partial y}$ and $\frac{\partial Q}{\partial x}$ exist and are continuous in D. Then the relation*

$$\frac{\partial P}{\partial y} = \frac{\partial Q}{\partial x} \tag{3.2}$$

is a necessary condition for ω to have a primitive in D; it is also sufficient if D is an open disc.

Proof. From formula (3.1), condition (3.2) implies that $\int_\gamma \omega = 0$ whenever γ is the boundary of a rectangle contained in D; if D is an open disc, this implies that ω has a primitive (proposition 2. 2). Conversely, if $\int_\gamma \omega = 0$ whenever γ is the boundary of a rectangle A contained in D with sides parallel to the axes, we have

$$\iint_A \left(\frac{\partial P}{\partial y} - \frac{\partial Q}{\partial x}\right) dx\, dy = 0 \tag{3.3}$$

for any such rectangle A. Moreover, this implies relation (3. 2). For, if the continuous function $\frac{\partial P}{\partial y} - \frac{\partial Q}{\partial x}$ is not identically zero in D, there will be some point of D in a neighbourhood of which it is > 0, say, and consequently the integral

$$\iint_A \left(\frac{\partial P}{\partial y} - \frac{\partial Q}{\partial x}\right) dx\, dy$$

will also be > 0 for a rectangle A contained in this neighbourhood, contrary to hypothesis (3. 3). Proposition 3. 3 is thus proved.

4. Closed differential forms

Definition. We say that a form $\omega = P\,dx + Q\,dy$, with continuous coefficients P and Q in an open set D, is *closed* if any point $(x_0, y_0) \in D$ has an open neighbourhood in which ω has a primitive. We can assume that such a neighbourhood is a disc with centre (x_0, y_0). Therefore, the results of nos. 2 and 3 immediately imply :

Proposition 4. 1. *A necessary and sufficient condition for a differential form ω with continuous coefficients in D to be closed is that $\int_\gamma \omega = 0$ whenever γ is the boundary of a small rectangle contained (with its interior) in D with sides parallel to the axes. If we also assume that P and Q have continuous partial derivatives of the first order, then (3. 2) is a necessary and sufficient condition for ω to be closed.*

We know from proposition 2. 2 that any closed form in an *open disc* has a primitive. We shall now give an example of a closed form ω in a connected open set D which has no primitive in D.

Proposition 4. 2. *Let D be the open set consisting of all points $z \neq 0$ of the complex plane **C**. The form $\omega = dz/z$ is closed in D but has no primitive.*

For, in a neighbourhood of each point $z_0 \neq 0$, there is a branch of $\log z$ and this branch is, in the neighbourhood of z_0, a primitive of dz/z. Hence ω is closed. To show that ω has no primitive in D, it is sufficient to find a closed path γ in D such that $\int_\gamma \frac{dz}{z} \neq 0$. In fact, let γ be the unit circle centred at the origin and described in the positive sense. To calculate $\int_\gamma \omega$, we put $z = e^{it}$ with t running from 0 to 2π; we have

$$dz = ie^{it}\,dt, \qquad \frac{dz}{z} = i\,dt,$$

and consequently

$$\int_\gamma \frac{dz}{z} = \int_0^{2\pi} i\,dt = 2i\pi \neq 0. \tag{4. 1}$$

This completes the proof.

In the preceding example, the form ω is complex. Let us now take the imaginary part of ω. Since

$$\frac{dz}{z} = \frac{dx + i\,dy}{x + iy} = \frac{x\,dx + y\,dy}{x^2 + y^2} + i\frac{x\,dy - y\,dx}{x^2 + y^2},$$

the differential form

$$\varpi = \frac{x\,dy - y\,dx}{x^2 + y^2}$$

is closed in the plane with the origin excluded. It has no primitive because we have by (4. 1)

$$\int_\gamma \frac{x\,dy - y\,dx}{x^2 + y^2} = 2\pi$$

if γ is the unit circle described in the positive sense. In fact, ϖ is the differential of arc tan $\frac{y}{x}$, which is a *many-valued* function (that is to say with many branches) in the plane with the origin excluded.

5. Study of many-valued primitives

Let ω be a closed form defined in a connected open set D. Although ω has not necessarily a (single-valued) primitive in D, we shall define what is meant by a *primitive of* ω *along a path* γ of D. Such a path is defined by a *continuous* mapping of the segment $I = [a, b]$ into D; we do not assume differentiability in this context.

Definition. Let $\gamma : [a, b] \to D$ be a path contained in an open set D, and let ω be a closed differential form in D. A continuous function $f(t)$ (t describing $[a, b]$) is called a *primitive of* ω *along* γ if it satisfies the following condition :
(P) *for any* $\tau \in [a, b]$ *there exists primitive* F *of* ω *in a neighbourhood of the point* $\gamma(\tau) \in D$ *such that*

(5. 1) $$F(\gamma(t)) = f(t)$$

for t *near enough to* τ.

THEOREM 1. *Such a primitive* f *always exists and is unique up to addition of a constant.*

Proof. First of all, if f_1 and f_2 are two such primitives, the difference $f_1(t) - f_2(t)$ is, by (5. 1), of the form $F_1(\gamma(t)) - F_2(\gamma(t))$ in a neighbourhood of each $\tau \in [a, b]$; since the difference $F_1 - F_2$ of two primitives of ω is constant, it follows that the function $f_1(t) - f_2(t)$ is constant in a neighbourhood of each point of the segment I. We express this by saying that the function $f_1 - f_2$ is *locally constant.* However, a continuous locally constant function on a *connected* topological space (the segment $I = [a, b]$ in this case) is *constant.* Indeed, for any number u, the set of points of the space where the function takes the value u is both open and closed.
It remains to be proved that there exists a continuous function $f(t)$ satisfying conditions (P). Each point $\tau \in I$ has a neighbourhood (in I)

mapped by γ into an open disc where ω has a primitive F. Since I is compact, we can find a finite sequence of points

$$a = t_0 < t_1 < \cdots < t_n < t_{n+1} = b,$$

such that, for each integer i where $0 \leqslant i \leqslant n$, γ maps the segment $[t_i, t_{i+1}]$ into an open disc U_i in which ω has a primitive F_i. The intersection $U_i \cap U_{i+1}$ contains $\gamma(t_{i+1})$ so it is not empty; it is connected, so $F_{i+1} - F_i$ is constant in $U_i \cap U_{i+1}$. We can then, by adding a suitable constant to each F_i, arrange, step by step, that F_{i+1} coincides with F_i in $U_i \cap U_{i+1}$. Then, we let $f(t)$ be the function defined by

$$f(t) = F_i(\gamma(t)) \qquad \text{for} \qquad t \in [t_i, t_{i+1}].$$

It is obvious that $f(t)$ is continuous and satisfies condition (P); the latter is clear when τ is different from the t_i and the reader should verify it when τ is equal to one of them.

Note. Suppose that γ is piecewise differentiable, in other words, that there is a subdivision of I such that the restriction of γ to each sub-interval $[t_i, t_{i+1}]$ is continuously differentiable. Then the integral $\int_\gamma \omega$ is defined; it is by definition

$$\sum_i \left(\int_{\gamma_i} \omega \right).$$

If f is a primitive along γ, we have by formula (2. 1)

$$\int_{\gamma_i} \omega = f(t_{i+1}) - f(t_i),$$

whence, by addition,

$$\int_\gamma \omega = f(b) - f(a). \tag{5. 2}$$

This leads to a definition of $\int_\gamma \omega$ for a *continuous* path γ, without the hypothesis of differentiability of γ : we take relation (5. 2) as the *definition*, which is valid because the right hand side does not depend on the choice of primitive f along γ.

PROPOSITION 5. 1 *If γ is a closed path which does not pass through the origin, $\frac{1}{2\pi i}\int_\gamma \frac{dz}{z}$ is an integer.*

Proof. $\omega = \dfrac{dz}{z}$ is a closed form. In the proof of theorem 1, we may suppose each F_i to be a branch of $\log z$. Thus $f(b) - f(a)$ is the difference between two branches of $\log z$ at the point $\gamma(a) = \gamma(b)$, and, consequently, is of the form $2\pi in$, where n is an integer.

COROLLARY. $\dfrac{1}{2\pi}\displaystyle\int_\gamma \frac{x\,dy - y\,dx}{x^2 + y^2}$ *is an integer* (the same integer as above).

The quantity $\displaystyle\int_\gamma \frac{x\,dy - y\,dx}{x^2 + y^2}$ is often called the *variation of the argument* of the point $z = x + iy$ when this point describes the path γ (whether γ is closed or not).

6. HOMOTOPY

For simplification, we shall only consider paths parametrized by the segment $I = [0, 1]$.

Definition. We say that two paths

$$\gamma_0 : I \to D \qquad \text{and} \qquad \gamma_1 : I \to D$$

having the same initial points and the same end points (that is to say $\gamma_0(0) = \gamma_1(0)$, $\gamma_0(1) = \gamma_1(1)$) *are homotopic* (in D) *with fixed end points*, if there exists a continuous mapping $(t, u) \to \delta(t, u)$ of $I \times I$ into D, such that

$$(6.1) \qquad \begin{cases} \delta(t, 0) = \gamma_0(t), & \delta(t, 1) = \gamma_1(t), \\ \delta(0, u) = \gamma_0(0) = \gamma_1(0), & \delta(1, u) = \gamma_0(1) = \gamma_1(1). \end{cases}$$

For fixed u, the mapping $t \to \delta(t,u)$ is a path γ_u of D with the same initial point as the common initial point of γ_0 and γ_1 and the same end point as their common end point. Intuitively, this path deforms continuously as u varies from 0 to 1, its end points remaining fixed.
There is an analogous definition for two *closed paths* γ_0 and γ_1 : we say that they are homotopic (in D) *as closed paths* if there is a continous mapping $(t, u) \to \delta(t, u)$ of $I \times I$ into D, such that

$$(6.2) \qquad \begin{cases} \delta(t, 0) = \gamma_0(t), & \delta(t, 1) = \gamma_1(t), \\ \delta(0, u) = \delta(1, u) & \text{for all } u, \end{cases}$$

(thus the path γ_u is closed for each u). In particular, we say that a closed path γ_0 is *homotopic to a point* in D if the above holds with $\gamma_1(t)$ a constant function.

THEOREM 2. *If γ_0 and γ_1 are two homotopic paths of* D *with fixed end points, then*

$$\int_{\gamma_0} \omega = \int_{\gamma_1} \omega$$

for any closed form ω *in* D.

THEOREM 2′. *If γ_0 and γ_1 are closed paths which are homotopic as closed paths then*

$$\int_{\gamma_0} \omega = \int_{\gamma_1} \omega$$

for any closed form ω.

These two theorems are consequences of a lemma which we shall now state. First of all, here is a definition :

Definition. Let $(t, u) \rightarrow \delta(t, u)$ be a continuous mapping of a rectangle

$$(6.3) \qquad a \leqslant t \leqslant b, \qquad a' \leqslant u \leqslant b'$$

into the open set D, and let ω be a closed form in D. A *primitive of* ω *following the mapping* δ is a *continuous* function $f(t, u)$ in the rectangle satisfying the following condition :
(P′) *For any point* (τ, υ) *of the rectangle, there exists a primitive* F *of* ω *in a neighbourhood of* $\delta(\tau, \upsilon)$ *such that*

$$F(\delta(t, u)) = f(t, u)$$

at any point (t, u) *sufficiently near to* (τ, υ).

LEMMA. *Such a primitive always exists and is unique up to addition of a constant,* This lemma is, in some sense, an extension of theorem 1. We shall prove it in an similar way. By using the compactness of the rectangle, we can quadrisect it by subdividing the interval of variation of t by points t_i and that of u by points u_j, in such a way that, for all i, j, the small rectangle, which is the product of the segments $[t_i, t_{i+1}]$, $[u_j, u_{j+1}]$, is mapped by δ into an open disc $U_{i,j}$, in which ω has a primitive $F_{i,j}$.

Keep j fixed; since the intersection $U_{i,j} \cap U_{i+1,j}$ is non-empty (and connected), we can add a constant to each $F_{i,j}$ (j fixed and i variable) in such a way that $F_{i,j}$ and $F_{i+1,j}$ coincide in $U_{i,j} \cap U_{i+1,j}$; we then obtain, for $u \in [u_j, u_{j+1}]$, a function $f_j(t, u)$ such that, for all i, we have

$$f_j(t, u) = F_{i,j}(\delta(t, u)) \qquad \text{when} \qquad t \in [t_i, t_{i+1}].$$

Hence $f_j(t, u)$ is continuous in the rectangle

$$a \leqslant t \leqslant b, \qquad u_j \leqslant u \leqslant u_{j+1},$$

and it is a primitive of ω following the mapping δ_j, the restriction of δ to this rectangle. Each function f_j is defined up to the addition of a constant; we can therefore, by induction on j, choose these additive constants in such a way that the functions $f_j(t, u)$ and $f_{j+1}(t, u)$ are equal when $u = u_{j+1}$. Finally, let $f(t, u)$ be the function defined in the rectangle (6. 3) by the condition that, for all j, we have

$$f(t, u) = f_j(t, u) \qquad \text{when} \qquad u \in [u_j, u_{j+1}].$$

This is a continuous function which satisfies conditions (P′) and is indeed a primitive of ω following the mapping δ. The lemma is thus proved.

Proof of theorem 2. Let δ be a continuous mapping satisfying conditions (6. 1) and let f be a primitive of ω following δ. It is obvious that f is a constant on the vertical sides $t = 0$ and $t = 1$ of the rectangle $I \times I$. Thus we have

$$f(0, 0) = f(0, 1), \qquad f(1, 0) = f(1, 1)$$

and, since

$$\int_{\gamma_0} \omega = f(1, 0) - f(0, 0), \qquad \int_{\gamma_1} \omega = f(1, 1) - f(0, 1),$$

theorem 2 is proved.

The proof of theorem 2′ is completely analogous; one uses a mapping δ satisfying (6. 2).

7. Primitives in a simply connected open set

Definition. We say that D is *simply connected* if it is connected and if in addition any closed path in D is homotopic to a point in D.

THEOREM. 3. *Any closed differential form ω in a simply connected open set* D *has a primitive in* D.

For, from theorem 2′, we have $\int_\gamma \omega = 0$ for any closed path γ contained in D, which implies by proposition 2. 1 that ω has a primitive in D.

In particular, in any simply connected open set not containing 0, the closed form dz/z has a primitive; in other words, log z *has a branch in any simply connected open set which does not contain* 0.

Examples of simply connected open sets. We say that a subset E of the plane is *starred* with respect to one of its points a if, for any point $z \in E$, the line segment joining a to z lies in E.

Any open set D *which is starred with respect to one of its points a is simply connected* : for, D is obviously connected; moreover, for each real number u between 0 and 1, the homothety of centre a and factor u transforms D into itself; as u decreases from 1 to 0, this homothety defines a homotopy of any closed curve to a point.

In particular, a *convex* open set D is *simply connected.* For, a convex open set is starred with respect to any of its points.

In contrast, the plane with the origin excluded is *not* simply connected : for example, the circle $|z| = 1$ is not homotopic to a point in $\mathbf{C} - \{0\}$ since the integral $\int \frac{dz}{z}$ of the closed form $\frac{dz}{z}$ along this circle is not zero (cf. relation (4.1)).

The reader is invited to prove the equivalence of the following four properties (for a connected open set D) as an exercise :

a) D is simply connected;

b) any continuous mapping of the circle $|z| = 1$ into D can be extended to a continuous mapping of the disc $|z| \leqslant 1$ into D;

c) any continuous mapping of the boundary of a square into D can be extended to a continuous mapping of the square itself into D.

d) if two paths of D have the same end points, then they are homotopic with fixed end points.

8. The index of a closed path

Definition. Let γ be a closed path in the plane $\mathbf{C}$ and let a be a point of $\mathbf{C}$ which does not belong to the image of γ. The *index* of γ with respect to a, denoted by $I(\gamma, a)$, is defined to be the value of the integral

$$\frac{1}{2\pi i}\int_\gamma \frac{dz}{z-a}. \tag{8.1}$$

Proposition 5.1 gives that the index $I(\gamma, a)$ is an *integer.* By referring back to the definitions, we see that, in order to calculate the index, we must find a continuous complex-valued function $f(t)$ defined for $0 \leqslant t \leqslant 1$ and such that

$$e^{f(t)} = \gamma(t) - a;$$

then we have

$$I(\gamma, a) = \frac{f(1) - f(0)}{2\pi i}.$$

Properties of the index

1) *If the point a is fixed, the index* $I(\gamma, a)$ *remains constant when the closed path* γ *is continuously deformed without passing through the point a.* This follows directly from theorem 2′ of no. 6.

2) *If the closed path* γ *is fixed, the index* $I(\gamma, a)$ *is a locally constant function of* a *when* a *varies in the complement of the image of* γ. The proof is the same as for 1). It follows that $I(\gamma, a)$ is a function of a which is *constant* in each connected component of the complement of the image of γ.

3) *If the image of* γ *is contained in a simply connected open set* D *which does not contain the point* a, *then the index* $I(\gamma, a)$ *is zero.* For, the closed path γ can then be deformed to a point while remaining in D, thus it never passes through a; it is sufficient now, to use 1).

4) *If* γ *is a circle described in the positive sense* (i.e. *in the sense such that* $I(\gamma, 0) = +1$), *the index* $I(\gamma, a)$ *is equal to* 0 *for* a *outside the circle and equal to* 1 *for* a *inside the circle.* The case when a is outside the circle is covered by 3); when a is inside the circle, it is sufficient to examine the case where a is the centre of the circle because of 2); so, we apply relation (4. 1).

PROPOSITION 8. 1. *Let* f *be a continuous mapping of the closed disc* $x^2 + y^2 \leqslant r^2$ *into the plane* $\mathbf{R}^2$ *and let* γ *be the restriction of* f *to the circle* $x^2 + y^2 = r^2$. *If a point* a *of the plane does not belong to the image of* γ *and if the index* $I(\gamma, a)$ *is* $\neq 0$, *then* f *takes the value* a *at least once in the open disc* $x^2 + y^2 < r^2$.

We prove this by *reductio ad absurdum* supposing that f does not take the value a. The restriction of f to concentric circles of centre 0 defines a continuous deformation of the closed path γ to a point. Consequently, the integral $\int_\gamma \frac{dz}{z - a}$ is zero, which contradicts the hypothesis.

Definition. Let γ_1 and γ_2 be two closed paths which do not pass through the origin 0. The *product* of these two paths means the closed path defined by the mapping

$$t \to \gamma_1(t) . \gamma_2(t),$$

where the dot means multiplication of the complex numbers $\gamma_1(t)$ and $\gamma_2(t)$.

PROPOSITION 8. 2. *The index, with respect to the origin, of the product of two closed paths, which do not pass through* 0, *is equal to the sum of the indices of each of these closed paths.* In other words,

$$I(\gamma_1\gamma_2, 0) = I(\gamma_1, 0) + I(\gamma_2, 0).$$

For, let $f_1(t)$ and $f_2(t)$ be two f continuous complex-valued functions such that

$$e^{f_1(t)} = \gamma_1(t), \qquad e^{f_2(t)} = \gamma_2(t).$$

Let $\gamma(t) = \gamma_1(t).\gamma_2(t)$ be the product of the two closed curves; the function $f(t) = f_1(t) + f_2(t)$ satisfies

$$e^{f(t)} = \gamma(t)$$

and we have

$$I(\gamma, 0) = \frac{f(1)-f(0)}{2\pi i} = \frac{f_1(1)-f_1(0)}{2\pi i} + \frac{f_2(1)-f_2(0)}{2\pi i} = I(\gamma_1, 0) + I(\gamma_2, 0),$$

which completes the proof.

PROPOSITION 8. 3. *Let γ and γ_1 be two closed paths in the plane* **C**. *If γ never takes the value* 0 *and if we always have $|\gamma_1(t)| < |\gamma(t)|$, then the mapping $t \to \gamma(t) + \gamma_1(t)$ never takes the value* 0 *and*

$$I(\gamma + \gamma_1, 0) = I(\gamma, 0).$$

For, we can write

$$\gamma(t) + \gamma_1(t) = \gamma(t).\left(1 + \frac{\gamma_1(t)}{\gamma(t)}\right);$$

the closed path $t \to 1 + \dfrac{\gamma_1(t)}{\gamma(t)}$ has zero index with respect to the origin because it is contained in the open disc of centre 1 and radius 1. Thus the closed path $\gamma + \gamma_1$ is the product of two closed paths γ and $1 + \dfrac{\gamma_1}{\gamma}$, and by applying proposition 8. 2, we obtain proposition 8.3.

9. COMPLEMENTS : ORIENTED BOUNDARY OF A COMPACT SET

LEMMA. *If a path γ is continuously differentiable and if its derivative γ' is everywhere $\neq 0$, then, in a neighbourhood of each value of the parameter t, the mapping $t \to \gamma(t)$ is injective and its image cuts the plane (locally) into two regions.*

The exact meaning of this statement will be made clear in the proof which follows. Let $t \to \gamma(t)$ be a continuously differentiable mapping of the segment $[a, b]$ into the plane $\mathbf{R}^2$ and let the derivative $\gamma'(t)$ be $\neq 0$ for all values of t. The coordinates x, y of the point $\gamma(t)$ are then continuously differentiable functions $\gamma_1(t)$, $\gamma_2(t)$ and their derivatives $\gamma_1'(t)$, $\gamma_2'(t)$ do not vanish simultaneously. The implicit function theorem shows then that, if t_0 is an interior point of the interval (that is, if $a < t_0 < b$) and if we write $x_0 = \gamma_1(t_0)$, $y_0 = \gamma_2(t_0)$, there exists a continuously differentiable mapping $(t, \omega) \to \delta(t, u)$ of an open neighbourhood U of the point $(t_0, 0)$ onto an open neighbourhood V of point (x_0, y_0), which satisfies the following conditions :

(i) $\delta(t, 0) = \gamma(t)$;

(ii) δ is a homeomorphism of U on V whose Jacobian is >0 at each point of U (thus δ preserves ' orientation '). Thus V is mapped homeomorphically by the inverse homeomorphism of δ onto U, the points of the path γ going onto the points of the line $u = 0$. The points of V complementary to γ are then partitioned into two open sets V^+ and V^-: that

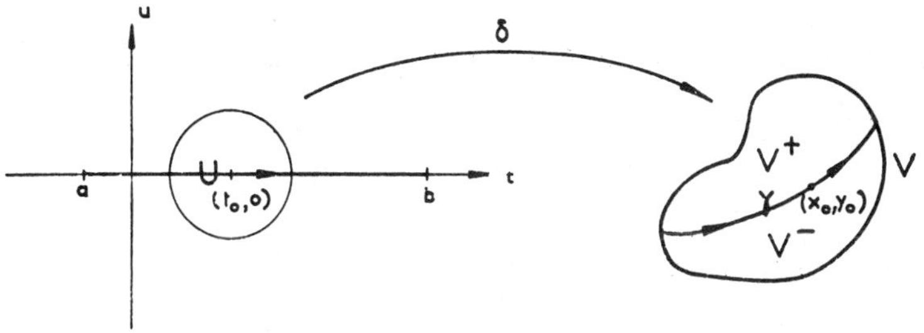

Fig. 2.

for which u is > 0 and that for which u is < 0. If we take U to be an open disc of centre $(t_0, 0)$, then the open sets V^+ and V^- are connected. Thus the path γ splits the open set V into two connected components, which completes the proof of the lemma.

Definition. Let K be a compact subset of the plane $\mathbf{C}$, and let $\Gamma = \{\Gamma_i\}_{i \in I}$ be a finite set of *closed* piecewise differentiable paths. We say that Γ is the *oriented boundary* of the compact set K if the following conditions are satisfied:

(BO 1) each mapping $t \to \Gamma_i(t)$ takes any two distinct points into distinct points, except for the initial and final points of the defining segment, and, moreover, the images of the various Γ_i are disjoint and their union is the *frontier* of K;

(BO 2) if γ is a differentiable path of any of the Γ_i, its derivative $\gamma'(t)$ is always $\neq 0$, and, if t_0 is an interior point of the defining interval of γ and the open set V of the previous lemma is chosen to be sufficiently small, then V^- *does not meet* K *while* V^+ *is contained in the interior of* K.
Condition (BO2) is expressed intuitively by saying that, when γ is described in the direction of t increasing, the interior points of K are always *on the left*, whereas the points in the complement of K are on the right.

Example. Take K to be a (closed) rectangle whose sides are parallel to the axes, then the perimeter of this rectangle, as defined at the end of n° 1, is the *oriented boundary of* K.

We shall admit, without proof, that the Green-Riemann formula holds for the oriented boundary Γ of a compact set K. A precise statement of the formula is that, if $\omega = P\,dx + Q\,dy$ is a differential form with continuously differentiable coefficients in an open set containing the compact set K, then

$$(9.1) \qquad \int_\Gamma P\,dx + Q\,dy = \iint_K \left(\frac{\partial Q}{\partial x} - \frac{\partial P}{\partial y}\right) dx\,dy.$$

$\left(\text{The notation } \int_\Gamma \text{ means } \sum_i \int_{\Gamma_i} \text{ where } \Gamma_i \text{ are the closed paths of } \Gamma\right).$

In particular, *if the form* ω *is closed in* D, *we have the relation*

$$(9.2) \qquad \int_\Gamma \omega = 0$$

whenever Γ *is the oriented boundary of a compact subset of* D.

2. Holomorphic Functions; Fundamental Theorems

1. Revision of differentiable functions

Let D be an open set of the plane $\mathbf{R}^2$ and let $f(x, y)$ be a real- or complex-valued function defined in D. We say that f is *differentiable* at the point $(x_0, y_0) \in D$ if there is a linear function $ah + bk$ of the real variables h and k, such that

$$(1.1) \qquad f(x_0 + h, y_0 + k) - f(x_0 y_0) = ah + bk + \alpha\sqrt{h^2 + k^2},$$

for all sufficiently small values of h and k; α is a (real-or complex-valued) function of h and k whose absolute value tends to 0 when $\sqrt{h^2 + k^2}$ tends to 0. If f is differentiable at the point (x_0, y_0), the (real or complex) constants a and b are uniquely determined and are equal to the partial derivatives

$$a = \frac{\partial f}{\partial x}(x_0, y_0), \qquad b = \frac{\partial f}{\partial y}(x_0, y_0).$$

Recall that the existence of the partial derivatives of f at the point (x_0, y_0) is not sufficient for the function to be differentiable at this point; but if f has partial derivatives at every point sufficiently near to (x_0, y_0) and if these partial derivatives are continuous functions at the point (x_0, y_0), then f is differentiable at this point. A function which has continuous partial derivatives in an open set D is said to be continuously differentiable in D.

2. Condition for holomorphy

Let D be an open subset of the complex plane $\mathbf{C}$ and let f be a function of the complex variable $z = x + iy$ defined in D.

Definition. We say that $f(z)$ is *holomorphic* at the point $z_0 \in D$ if

$$\lim_{\substack{u \to 0 \\ u \neq 0}} \frac{f(z_0 + u) - f(z_0)}{u} \quad \text{exists} \tag{2. 1}$$

(u denotes a variable complex number). This is the same as saying that f has a *derivative* with respect to the complex variable at the point z_0. We say that f is holomorphic in the open set D if it holomorphic at each point of D.

Condition (2. 1) can also be written

$$f(z_0 + u) - f(z_0) = cu + \alpha(u)\,|u|, \tag{2. 2}$$

where $\alpha(u)$ tends to 0 as u tends to 0; c is the derivative $f'(z_0)$. Since $z = x + iy$, relation (2. 2) can also be written

$$f(x_0 + h, y_0 + k) - f(x_0, y_0) = c(h + ik) + \alpha(h, k)\sqrt{h^2 + k^2}. \tag{2. 3}$$

This shows that f, considered as a function of two real variables x and y, is differentiable and that

$$a = c, \qquad b = ic,$$

where a and b are the constants in relation (1. 1). Thus we have $\frac{\partial f}{\partial x} = c$, and $\frac{\partial f}{\partial y} = ic$, whence

$$\frac{\partial f}{\partial x} + i\frac{\partial f}{\partial y} = 0. \tag{2. 4}$$

Conversely, let f be a differentiable function of the real variables x and y satisfying (2. 4). Then, relation (1. 1) implies (2. 3) with $c = a$ and $ic = b$. Thus, f is holomorphic at the point $z_0 = x_0 + iy_0$. We have, in fact, proved the following proposition :

PROPOSITION 2. 1. *For f to be holomorphic at a point, it is necessary and sufficient that f, considered as a function of the real variables x and y, is differentiable at this point and that relation* (2. 4) *holds between the partial derivatives of f at this point.*

We express relation (2. 4) more explicitly : if we put $f = P + iQ$, where P and Q are real functions, then we obtain the Cauchy conditions

$$\frac{\partial P}{\partial x} = \frac{\partial Q}{\partial y}, \qquad \frac{\partial P}{\partial y} = -\frac{\partial Q}{\partial x}. \tag{2. 5}$$

3. Introduction of the variables z and $\bar{z}$

Let f be a (real- or complex-valued) differentiable function of the real variables x and y. Consider the differential

$$df = \frac{\partial f}{\partial x}\, dx + \frac{\partial f}{\partial y}\, dy. \tag{3.1}$$

The particular functions $z = x + iy$ and $\bar{z} = x - iy$ have differentials

$$dz = dx + i\, dy, \qquad d\bar{z} = dx - i\, dy; \tag{3.2}$$

thus we have conversely

$$dx = \frac{1}{2}(dz + d\bar{z}), \qquad dy = \frac{1}{2i}(dz - d\bar{z}). \tag{3.3}$$

By substituting this in (3. 1) we obtain the equation

$$df = \frac{1}{2}\left(\frac{\partial f}{\partial x} - i\frac{\partial f}{\partial y}\right) dz + \frac{1}{2}\left(\frac{\partial f}{\partial x} + i\frac{\partial f}{\partial y}\right) d\bar{z}.$$

This leads us to introduce the symbols

$$\frac{\partial}{\partial z} = \frac{1}{2}\left(\frac{\partial}{\partial x} - i\frac{\partial}{\partial y}\right), \qquad \frac{\partial}{\partial \bar{z}} = \frac{1}{2}\left(\frac{\partial}{\partial x} + i\frac{\partial}{\partial y}\right). \tag{3.4}$$

With this notation, we obtain the equation

$$df = \frac{\partial f}{\partial z}\, dz + \frac{\partial f}{\partial \bar{z}}\, d\bar{z}. \tag{3.5}$$

Condition (2. 4), which expresses that f is a holomorphic function of the complex variable z, can now be written

$$\frac{\partial f}{\partial \bar{z}} = 0. \tag{3.6}$$

In other words, a necessary and sufficient condition for f to be holomorphic is that the coefficient of $d\bar{z}$ is zero in the expression (3. 5) for the differential df. Or again: df must be proportional to dz, the coefficient of proportionality being simply the derivative $f'(z)$.

We shall apply this to prove the following result : *Let f be a holomorphic function in a connected open set* D; *if the real part of f is constant, then f is constant.*

For, the real part $\mathrm{Re}(f)$ is simply $\frac{1}{2}(f + \bar{f})$; by hypothesis $d(f + \bar{f}) = 0$ in D, which can be written

$$\frac{\partial f}{\partial z}\, dz + \frac{\partial f}{\partial \bar{z}}\, d\bar{z} + \frac{\partial \bar{f}}{\partial z}\, dz + \frac{\partial \bar{f}}{\partial \bar{z}}\, d\bar{z} = 0.$$

But, since f is is holomorphic, we have $\frac{\partial f}{\partial \bar{z}} = 0$; by passing to the complex conjugate, we have $\frac{\partial \bar{f}}{\partial z} = 0$. Hence,

$$\frac{\partial f}{\partial z} dz + \frac{\partial \bar{f}}{\partial \bar{z}} d\bar{z} = 0.$$

However, an expression $adz + bd\bar{z}$ can only be identically zero if the coefficients a and b are zero, which gives $\frac{\partial f}{\partial z} = 0$, $\frac{\partial \bar{f}}{\partial \bar{z}} = 0$. Thus, $df = 0$, and f is constant in D.

We deduce from this that, *if f is holomorphic and $\neq 0$ in a connected open set* D *and if either* $\log |f|$ *is constant or* $\arg f$ *is constant, then f is constant.*

For, consider the function

$$g(z) = \log f(z) = \log |f(z)| + i \arg f(z).$$

We stay in some neighbourhood of the point z_0 and we choose a branch of the argument; g is holomorphic and its real (or imaginary) part is constant. Thus g is constant in some neighbourhood of z_0. Thus $f = e^g$ is locally constant in D and is consequently constant since D is connected.

4. Cauchy's theorem

Theorem 1. *If $f(z)$ is holomorphic in an open set* D *of the complex plane, then the differential form $f(z)dz$ is closed in* D.

In view of the importance of this theorem, we shall give two proofs :

First proof. This proof requires an extra hypothesis. We suppose that the partial derivatives $\frac{\partial f}{\partial x}$ and $\frac{\partial f}{\partial y}$ are continuous in D. (In fact the second proof shows that this hypothesis is automatically satisfied when f is holomorphic.) To verify that the differential form $f(z)\, dz = f(z)\, dx + if(z)\, dy$ is closed, it is sufficient, by the Green-Riemann formula (§ 1, formula (3. 1)), to verify that

$$\frac{\partial f}{\partial y} = i \frac{\partial f}{\partial x}.$$

However, this is precisely condition (2. 4) expressing that f is holomorphic, and the proof is completed.

Second proof. This proof, unlike the first, does not need any additional hypothesis, but it requires a more subtle argument. To show that $f(z)dz$

is closed, we must prove that the integral $\int_\gamma f(z)\,dz$ is zero along the boundary γ of any rectangle R contained (with its interior) in D. To this end, we put *a priori*

$$\int_\gamma f(z)\,dz = \alpha(\mathrm{R}). \tag{4. 1}$$

Divide the rectangle R into four equal rectangles by dividing each side into two equal parts.

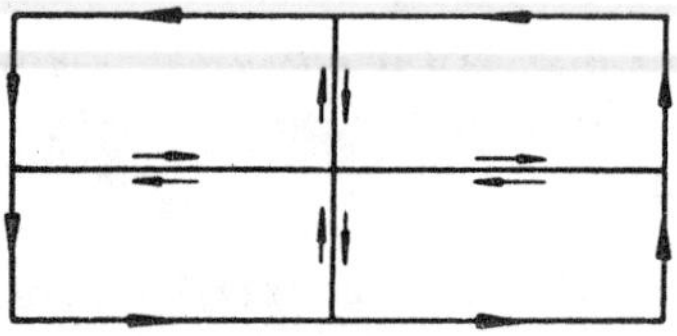

Fig. 3.

Let γ_i be the (oriented) boundaries of the four small rectangles ($i = 1, 2, 3, 4$) It is easily verified (cf. fig. 3) that

$$\int_\gamma f(z)\,dz = \sum_{i=1}^{4} \int_{\gamma_i} f(z)\,dz = \sum_{i=1}^{4} \alpha(\mathrm{R}_i).$$

Thus among there four rectangles there is at least one such that $|\alpha(\mathrm{R}_i)| \geqslant \frac{1}{4}|\alpha(\mathrm{R})|$. Call this rectangle $\mathrm{R}^{(1)}$. Now divide the rectangle $\mathrm{R}^{(1)}$ into four equal rectangles at least one of which, say $\mathrm{R}^{(2)}$, will satisfy the condition $|\alpha(\mathrm{R}^{(2)})| \geqslant \frac{1}{4^2}|\alpha(\mathrm{R})|$. We can repeat this operation indefinitely to obtain a sequence of rectangles each included in the previous one; the kth rectangle $\mathrm{R}^{(k)}$ will have sides 2^k times smaller than those of R and its area will then be 4^k times smaller than that of the rectangle R. If $\gamma(\mathrm{R}^{(k)})$ denotes the oriented boundary of the rectangle $\mathrm{R}^{(k)}$, then

$$\left|\int_{\gamma(\mathrm{R}^{(k)})} f(z)\,dz\right| \geqslant \frac{1}{4^k}|\alpha(\mathrm{R})|. \tag{4. 2}$$

By the Cauchy criterion of convergence, there is a unique point z_0 common to all the rectangles $\mathrm{R}^{(k)}$. Obviously $z_0 \in \mathrm{D}$. Thus $f(z)$ is holomorphic at the point z_0, and, consequently,

$$f(z) = f(z_0) + f'(z_0)\,(z - z_0) + \varepsilon(z)\,|z - z_0|$$

with

$$\lim_{z \to z_0} \varepsilon(z) = 0.$$

We deduce

$$(4.3)\quad \begin{cases} \int_{\gamma(R^{(k)})} f(z)\, dz = f(z_0) \int_{\gamma(R^{(k)})} dz + f'(z_0) \int_{\gamma(R^{(k)})} (z - z_0)\, dz \\ \qquad\qquad + \int_{\gamma(R^{(k)})} \varepsilon(z)\, |z - z_0|\, dz. \end{cases}$$

On the right hand side of (4. 3), the first two integrals are zero and the third is negligible compared with the area of the rectangle $R^{(k)}$ as k increases indefinitely; it is then negligible compared with $\frac{1}{4^k}$. Comparing this with (4. 2) shows that we must have $\alpha(R) = 0$; consequently by the definition of $\alpha(R)$, we have $\int_\gamma f(z)\, dz = 0$. This completes the proof.

Corollary 1. *A holomorphic function $f(z)$ in D has locally a primitive, which is holomorphic.*

This statement means that any point of D has an open neighbourhood in which f has a holomorphic primitive. The local existence of a primitive follows from the definition of a closed form; and this local primitive is indeed holomorphic because it has f as its derivative.

Corollary 2. *If $f(z)$ is holomorphic in D, then $\int_\gamma f(z)\, dz = 0$ for any closed path γ of D which is homotopic to a point in D.*

This follows from theorem 1 above and theorem 2′ of § 1, no. 6.

Generalization. We shall prove theorem 1 again with less strict conditions.

Theorem 1′. *Let $f(z)$ be an continuous function in a open set D, which is holomorphic at every point of D except perhaps at the points of a line Δ parallel to the real axis. Then the form $f(z)\, dz$ is closed. In particular, if f is holomorphic at any point of D except perhaps at some isolated points, then the form $f(z)\, dz$ is closed.*

Proof. We must prove that the integral $\int_\gamma f(z)\, dz$ is zero for the boundary γ of any rectangle contained in D. However, this is obvious if the rectangle does not intersect the line Δ. Suppose that the rectangle has a side contained in Δ and let $u, u + a, u + ib, u + a + ib$ be the four corners of the rectangle, u and $u + a$ being on the line Δ; a and b are real, and we assume, say, that $b > 0$. Let $R(\varepsilon)$ be the rectangle with corners

$$u + i\varepsilon, \quad u + a + i\varepsilon, \quad u + ib, \quad u + a + ib,$$

ε being a very small number > 0; the integral $\int f(z)\,dz$ is zero round the boundary of $R(\varepsilon)$; however, as ε tends to 0, this integral tends to the integral round the boundary γ of the rectangle R. Thus $\int_\gamma f(z)\,dz = 0$. Finally, if the line Δ meets the rectangle without containing one of its horizontal sides, the line Δ splits R into two rectangles R′ and R″ and the integral $\int f(z)\,dz$ is zero when taken round the boundaries of either R′ or R″, because of the previous remarks; however, the sum of these integrals is equal to the integral $\int f(z)\,dz$ round the boundary of R. This completes the proof.

5. Cauchy's integral formula

Theorem 2. *Let f be a holomorphic function in an open set D. Let $a \in D$ and let γ be a closed path of D which does not pass through a and which is homotopic to a point in D. Then,*

$$\frac{1}{2\pi i}\int_\gamma \frac{f(z)\,dz}{z-a} = I(\gamma, a)f(a), \tag{5.1}$$

where $I(\gamma, a)$ denotes the index of the closed path γ with respect ot a (cf. § 1, no. 8).

Proof. Let $g(z)$ be the function defined in D by

$$\begin{cases} g(z) = \dfrac{f(z)-f(a)}{z-a} & \text{for} \quad z \neq a, \\ g(z) = f'(a) & \text{for} \quad z = a; \end{cases}$$

this function g is continuous because of the definition of the derivative. It is holomorphic at any point of D except the point a. By theorem 1′, we have

$$\int_\gamma \frac{f(z)-f(a)}{z-a}\,dz = 0.$$

However,

$$\int_\gamma \frac{f(a)\,dz}{z-a} = 2\pi i\; I(\gamma, a)f(a),$$

by the definition of the index. This proves relation (5.1).

Example. Let f be a holomorphic function in some neighbourhood of a closed disc and let γ be the boundary of the disc described in the positive sense. Then,

$$\int_\gamma \frac{f(z)\,dz}{z-a} = \begin{cases} 2\pi i\,f(a) & \text{if } a \text{ is inside the disc,} \\ 0 & \text{if } a \text{ is outside the disc.} \end{cases}$$

6. Taylor expansion of a holomorphic function

THEOREM 3. *Let $f(z)$ be a holomorphic function in the open disc $|z| < \rho$; then f can be expanded as a power series in this disc.*

This means that there exists a power series $S(X) = \sum_{n \geqslant 0} a_n X^n$ whose radius of convergence is $\geqslant \rho$ and whose sum $S(z)$ is equal to $f(z)$ for $|z| < \rho$.

Proof. Let r be $< \rho$. We shall find a power series which converges normally to $f(z)$ for $|z| \leqslant r$. This series will be independent of r because of the uniqueness of the power series expansion of a function in a neighbourhood of 0. The theorem will then be proved.

Choose an r_0 such that $r < r_0 < \rho$. We shall apply the integral formula of theorem 2 by taking γ to be the circle of radius r_0 centred at 0 described in the positive sense :

$$f(z) = \frac{1}{2\pi i}\int_\gamma \frac{f(t)\,dt}{t-z} \qquad \text{for} \qquad |z| \leqslant r.$$

The function $\dfrac{1}{t-z}$ which occurs under the integral sign can be expanded as a series since $|z| < |t|$. Explicitly,

$$\frac{1}{t-z} = \frac{1}{t}\,\frac{1}{1-z/t} = \frac{1}{t}\left(1 + \frac{z}{t} + \cdots + \frac{z^n}{t^n} + \cdots\right);$$

consequently,

$$f(z) = \frac{1}{2\pi i}\int_\gamma \sum_{n \geqslant 0} z^n \frac{f(t)}{t^{n+1}}\,dt.$$

The series converges normally for $|z| \leqslant r$ and $|t| = r_0$. We can therefore integrate term by term and we obtain a normally convergent series for $|z| \leqslant r$:

$$f(z) = \sum_{n \geqslant 0} a_n z^n,$$

where the coefficients are given by the integrals

$$a_n = \frac{1}{2\pi i}\int_{|t|=r_0} \frac{f(t)\,dt}{t^{n+1}}. \tag{6.1}$$

Hence we have proved theorem 3.

Comment. Theorem 3 shows that any holomorphic function in an open set D is *analytic* in D. Conversely, any analytic function in D is holomorphic in D since we know that analytic functions have derivatives. Hence, for functions of a *complex* variable, there is an equivalence between *holo-*

morphy and *analyticity*. If we apply the known results for analytic functions to holomorphic functions, we see that *a holomorphic function is infinitely differentiable* and, in particular, is continuously differentiable, and that *the derivative of a holomorphic function is holomorphic.*

7. Morera's theorem

Theorem 4. (Converse of theorem 1). *Let $f(z)$ be a continuous function in an open set* D. *If the differential form $f(z)\,dz$ is closed, then the function $f(z)$ is holomorphic in* D.

For, f has a primitive g locally. This primitive is holomorphic, and $f = g'$ is the derivative of a holomorphic function, so is itself holomorphic from the above remarks.

Corollary. *If $f(z)$ is continuous in* D *and holomorphic at all points of* D *except perhaps at the points of some line Δ, then f is holomorphic at all points of* D *without exception.*

For, we can suppose Δ to be parallel to the real axis, by rotating if necessary. By theorem 1′, the form $f(z)\,dz$ is closed. Thus by theorem 4, f is holomorphic at all points of D.

We see then that theorem 1′ was only an apparent generalization of theorem 1. However, we needed to establish it for technical reasons.

8. Alternative form of Cauchy's integral formula

Theorem 5. *Let Γ be the oriented boundary of a compact subset* K *of an open set* D *and let $f(z)$ be a holomorphic function in* D. *Then,*

$$\int_\Gamma f(z)\,dz = 0;$$

if, moreover, a is an interior point of K, *then*

$$\int_\Gamma \frac{f(z)\,dz}{z-a} = 2\pi i f(a). \tag{8.1}$$

Proof. The first assertion follows from relation (9.1) of § 1. To prove the second assertion, we consider a small open disc S centred at a whose closure is in the interior of K. The oriented boundary of the compact set K — S is composed of Γ and the frontier-circle of S described in the negative sense. We shall say that this oriented boundary is the *difference* of Γ and the frontier-circle γ of S taken in the positive sense.

By applying the first part of theorem 5 to the compact set $K - S$ and the function $\frac{f(z)}{z-a}$, which is holomorphic in $D - \{a\}$, we obtain

$$\int_{\Gamma} \frac{f(z)\, dz}{z-a} = \int_{\gamma} \frac{f(z)\, dz}{z-a},$$

which, along with theorem 2, gives relation (8. 1).

9. Schwarz' principle of symmetry

We have seen (corollary to theorem 4) that, if $f(z)$ is continuous in an open set D and holomorphic at any point of D except perhaps at points on the real axis, then f is holomorphic at all points of D without exception. Consider, then, a non-empty, connected, open set D which is *symmetric with respect ot the real axis*; let D′ be the intersection of D with the closed half-plane $y \geqslant 0$ and let D″ be the intersection of D with the half-plane $y \leqslant 0$. Suppose we are given a function $f(z)$ which is continuous in D′, which takes real values at the points of the real axis, and which is holomorphic at points of D′ where $y > 0$. We shall show that *there is a holomorphic function in* D *which extends f;* such a function is unique by the principle of analytic continuation (cf. chap. 1, § 4, no. 3).

Consider the function $g(z)$ defined in D″ by the equation

$$g(z) = \overline{f(\bar{z})}.$$

This function is continuous in D″ and it can quickly be shown that it is holomorphic at any point of D″ not lying on the real axis. The function $h(z)$ which is equal to $f(z)$ in D′ and $g(z)$ in D″ is continuous in D and holomorphic at all points of D not lying on the real axis. It is therefore holomorphic at all points of D without exception.

Note that the function h takes complex conjugate values (that is, symmetric values with respect to the real axis) at pairs of points of D which are symmetric with respect to the real axis. This is why the preceding construction is called the " principle of symmetry ".

Exercises

1. a) Let γ be a piecewise differentiable path and let γ be its image under the mapping $z \to \bar{z}$ (symmetry with respect to the real axis.) Show that, if $f(z)$ is a continuous function on γ, the function $z \to \overline{f(\bar{z})}$ is continuous on $\bar{\gamma}$ and that

$$\overline{\int_{\gamma} f(z)\, dz} = \int_{\bar{\gamma}} \overline{f(\bar{z})}\, dz.$$

(b) In particular, if γ is the unit circle described in the positive sense, then

$$\overline{\int_\gamma f(z)\, dz} = -\int_\gamma \overline{f(z)}\, \frac{dz}{z^2}.$$

2. Let γ be a continuous path (not necessarily piecewise differentiable). Show that

$$\int_\gamma (\omega_1 + \omega_2) = \int_\gamma \omega_1 + \int_\gamma \omega_2,$$

$$\int_\gamma a\omega = a \int_\gamma \omega,$$

if ω_1, ω_2, ω are closed forms and $a \in \mathbf{C}$. $\left(\text{For the definition of } \int_\gamma \omega\text{, see}\right.$ *Note* § 1, no. 5.)

3. Let γ be a piecewise differentiable path, whose image is contained in an open set D, and let $\varphi(z)$ be a holomorphic function in D taking values in an open set Δ (of the plane of the complex variable w). Show that $\Gamma = \varphi \circ \gamma$ is a piecewise differentiable path and that, for any continuous function $f(\omega)$,

$$\int_\Gamma f(w)\, dw = \int_\gamma f(\varphi(z))\, \varphi'(z)\, dz.$$

Is this formula still true when γ is no longer necessarily differentiable?

4. Let γ be the (differentiable) path $t \to \gamma(t) = re^{it}$, $0 \leqslant t \leqslant 2\pi$, and let γ_n be the path $t \to \gamma_n(t) = (1 - 1/n)re^{it}$, with t varying over the same interval. Show that, if $f(z)$ is continuous in the closed disc $|z| \leqslant r$, then

$$\int_\gamma f(z)\, dz = \lim_{n\to\infty} \int_{\gamma_n} f(z)\, dz.$$

5. Show that, if $f(z)$ is *continuous* in the closed disc $|z| \leqslant r$ and holomorphic in the open disc $|z| < r$, then

$$f(z) = \frac{1}{2\pi i} \int_{|t|=r} \frac{f(t)}{t - z}\, dt \quad \text{for all} \quad |z| < r,$$

where the integral is taken in the positive sense.

6. Find a path $t \to \gamma(t)$ with t varying in $[0, 2\pi]$, having the ellipse $x^2/a^2 + y^2/b^2 = 1$ in the plane $\mathbf{R}^2$ $(a, b > 0)$ as image. Calculate the integral $\int_\gamma \frac{dz}{z}$ in two different ways, and deduce that

$$\int_0^{2\pi} \frac{dt}{a^2 \cos^2 t + b^2 \sin^2 t} = \frac{2\pi}{ab}.$$

7. Let $P_n(t) = t^n + a_{n-1}t^{n-1} + \cdots + a_0$ be a polynomial of degree $n \geqslant 1$ with complex coefficients and let γ_R be the image of the circle $|t| = R$ under the mapping $t \to z = P_n(t)$. Show that, if R is sufficiently large, γ_R does not pass through the origin $z = 0$ and that $I(\gamma_R, 0) = n$; deduce from this that $P_n(t) = 0$ has at least one root. (First show that, for sufficiently large R, $|t^n| > |a_{n-1}t^{n-1} + \cdots + a_0|$ for $|t| \geqslant R$. Then use proposition 8. 3 of § 1 to show that $I(\gamma_R, 0)$ is equal to the index, with respect to the origin, of the image of the circle $|t| = R$ by the mapping $t \to t^n$).

8. Let $f(z) = u(x, y) + iv(x, y)$ be a holomorphic function in a connected open set D. If

$$au(x, y) + bv(x, y) = c \quad \text{in} \quad D,$$

where a, b and c are real constants which are not all zero, then $f(z)$ is constant in D.

9. Let D be a *convex* open set in the plane and let $f(z)$ be a holomorphic function in D. Show that, for any pair of points $a, b \in D$, we can choose two points c and d on the line segment joining a and b such that

$$f(a) - f(b) = (a - b)\,(\mathrm{Re}\,(f'(c) + i\,\mathrm{Im}(f'(d))).$$

(Consider the function of a real variable r defined by

$$F(t) = f(b + (a - b)t)/(a - b),$$

and apply the mean value theorem to the real and imaginary parts of $F(t)$.)

10. Let D be a connected open set, which is symmetrical with respect to the real axis and has non-empty intersection I with it. Any holomorphic function $f(z)$ in D can be expreseed uniquely in the form

$$f(z) = g(z) + ih(z) \qquad \text{for all} \qquad z \in D,$$

where g and h are holomorphic functions in D which take *real* values in I. Show that, in this case,

$$\overline{g(\bar{z})} = g(z), \quad \overline{h(\bar{z})} = h(z)$$

and

$$\overline{f(\bar{z})} = g(z) - ih(z), \quad \text{for all} \quad z \in D.$$

11. Let f and g be two holomorphic functions in a connected open set D of the plane, which have no zeros in D; if there is a sequence (a_n) of points of D such that

$$\lim a_n = a, \quad a \in D \quad \text{and} \quad a_n \neq a \quad \text{for all } n,$$

and if

$$\frac{f'(a_n)}{f(a_n)} = \frac{g'(a_n)}{g(a_n)} \quad \text{for all } n,$$

show that there exists a constant c such that $f(z) = cg(z)$ in D.

12. Let $\varphi(z)$ be a *continuous* function on the oriented boundary Γ of a compact set K. Let D be open set complementary to Γ in $\mathbf{C}$, and put, for $z \in D$,

$$f(z) = \int_\Gamma \frac{\varphi(\zeta)}{\zeta - z}\, d\zeta.$$

(i) If $\rho = \inf_{\zeta \in \Gamma} |\zeta - a|$ for $a \in D$, show that $\frac{1}{\zeta - z}$, for $\zeta \in \Gamma$ and $|z - a| \leqslant r$ with $0 < r < \rho$, can be expanded in a series of powers of $(z - a)$ which is normally convergent; deduce that $f(z)$ is analytic in a neighbourhood of each $a \in D$. (cf. the proof of theorem 3, § 2.)

(ii) Show that

$$f^{(n)}(a) = n! \int_\Gamma \frac{\varphi(\zeta)}{(\zeta - a)^{n+1}}\, d\zeta,$$

for any integer $n \geqslant 1$, $a \in D$ (cf. chapter III, § 1).

13. Let $f(z)$ be holomorphic in $|z| < \rho$; show that, if $0 < r < \rho$, then

$$\lim_{\substack{h \to 0 \\ 0 < |h| < \rho - r}} \frac{f(z + h) - f(z)}{h} = f'(z)$$

uniformly for $|z| \leqslant r$. (By using 12., show that

$$\frac{f(z + h) - f(z)}{h} - f'(z) = \frac{h}{2\pi i} \int_{|t| = r'} \frac{f(t)\, dt}{(t - z - h)(t - z)^2},$$

where $r' = (\rho + r)/2$, $|h| < (r' - r)/2 = (\rho - r)/4$, say, and deduce from this that, if $M = \sup_{|t| = r'} |f(t)|$, then

$$\left| \frac{f(z + h) - f(z)}{h} - f'(z) \right| \leqslant 4M \frac{\rho + r}{(\rho - r)^3} |h|.)$$

14. If two closed paths of $\mathbf{C} - \{0\}$ have the same index with respect to 0, show that they are homotopic as closed paths in $\mathbf{C} - \{0\}$.

CHAPTER III

Taylor and Laurent Expansions. Singular Points and Residues

1. Cauchy's Inequalities; Liouville's Theorem

1. Integral formula for the Taylor coefficients

We have seen (chapter II, § 2, no. 6. theorem 3) that, if $f(z)$ is holomorphic in an open disc D centred at the origin, then $f(z)$ is the sum of a power series $\sum_{n \geqslant 0} a_n z^n$ which converges in D. The coefficients a_n of this power series are given by the relation

$$a_n = \frac{1}{n!} f^{(n)}(0).$$

In other words, the a_n are the coefficients of the Taylor expansion of $f(z)$ at the origin. This power series is called the Taylor series of $f(z)$. We now propose to express the coefficients a_n in terms of integrals involving the function f.

Put $z = re^{i\theta}$ for $0 \leqslant r < \rho$, where ρ denotes the radius of the disc D. We have

$$f(re^{i\theta}) = \sum_{n \geqslant 0} a_n r^n e^{in\theta}. \tag{1.1}$$

If we fix r, allowing θ to vary, $f(re^{i\theta})$ is a periodic function of θ, and the above relation gives the Fourier expansion of this function. We observe that, only the $e^{in\theta}$ occur in this expansion for the various integers $n \geqslant 0$. However, we know that the coefficients in the Fourier expansion of a continuous function of period 2π are expressible as integrals involving the function. In the present context, the series (1. 1) converges normally when θ varies, r remaining fixed; we can then integrate term by term and obtain

$$\frac{1}{2\pi}\int_0^{2\pi} e^{-in\theta} f(re^{i\theta})\, d\theta = \sum_{p\geqslant 0} \frac{1}{2\pi}\int_0^{2\pi} a_p r^p e^{i(p-n)\theta}\, d\theta;$$

on the right hand side, all the integrals are zero except that which corresponds to $p = n$, and we obtain the fundamental formula

$$a_n r^n = \frac{1}{2\pi}\int_0^{2\pi} e^{-in\theta} f(re^{i\theta})\, d\theta, \tag{1. 2}$$

which we could also have deduced from relation (6. 1) of chapter II, § 2. This integral formula gives an upper bound for the coefficient a_n : let $M(r)$ be the upper bound of $|f(re^{i\theta})|$ as θ varies, that is the upper bound of the values of f on the circumference of radius r. The absolute value of the right hand side of (1. 2) is then bounded above by $M(r)$, and relation (1. 2) thus gives the fundamental inequalities

$$|a_n| \leqslant \frac{M(r)}{r^n}, \qquad n \text{ an integer} \geqslant 0. \tag{1. 3}$$

These inequalities are known as the *Cauchy inequalities.*

2. Liouville's theorem

Theorem. *A bounded, holomorphic function $f(z)$ in the whole plane is constant.*

Proof. We apply inequality (1. 3) for any integer $n \geqslant 1$. The quantity $M(r)$ is, by hypothesis, less than some number M independent of r. Hence

$$|a_n| \leqslant \frac{M}{r^n}$$

no matter how big r is. Since the right hand side of this inequality tends to 0 as r tends to infinity (n being $\geqslant 1$), we see that $a_n = 0$ for $n \geqslant 1$, thus $f(z) = a_0$ is constant.

Application : d'Alembert's theorem. We shall show that any polynomial with complex coefficients which is not constant has at least one complex root. Let $P(z)$ be such a polynomial, we shall use *reductio ad absurdum* by supposing that $P(z) \neq 0$ for any complex number z. Then, the function $\frac{1}{P(z)}$ is holomorphic in the whole plane. It is bounded; for,

$$P(z) = a_n z^n + a_{n-1}z^{n-1} + \cdots + a_0 = z^n\left(a_n + \frac{a_{n-1}}{z} + \cdots + \frac{a_0}{z^n}\right), \qquad a_n \neq 0,$$

tends to infinity as $|z|$ tends to infinity, so there is a compact disc outside of which $\left|\frac{1}{P(z)}\right|$ is bounded; on the other hand, $\left|\frac{1}{P(z)}\right|$ is bounded in the compact disc because it is continuous function. Hence, $\frac{1}{P(z)}$ is bounded in the whole of the plane and so is constant by Liouville theorem. It follows that $P(z)$ is a constant, contrary to hypothesis.

2. Mean Value Property; Maximum Modulus Principle

1. Mean value property

We apply relation (1.2) of § 1 in the particular case when $n = 0$. Then,

$$a_0 = \frac{1}{2\pi}\int_0^{2\pi} f(re^{i\theta})\,d\theta,$$

or

$$f(0) = \frac{1}{2\pi}\int_0^{2\pi} f(re^{i\theta})\,d\theta. \tag{1.2}$$

This equation says that the value of f at the point 0 is equal to the mean value of f on the circle of centre 0 and radius r. It follows, more generally, that, if S is a closed disc contained in an open set D in which f is holomorphic, the value of f at the centre of S is equal to the mean of the values of f on the frontier circle of S (this mean being calculated with respect to the arc of the circle). We shall say that a real- or complex-valued, continuous function f defined in an open set D has the *mean value property* if, for any compact disc S contained in D, the value of f at the centre of S is equal to the mean value of f on the frontier circle of S. We shall see later that the functions with the mean value property are precisely the *harmonic functions*. From now on, we can say that *any holomorphic function has the mean value property*. It is clear that, if a complex-valued function has the mean value property, then so have its real and imaginary parts. Thus, the real and imaginary parts of a holomorphic function have the mean value property.

2. Maximum modulus principle

This principle will apply to any (real- or complex-valued) function which has the mean value property (that is to say, as we shall see later, to any harmonic function).

THEOREM 1. (maximum modulus principle). *Let f be a continuous (complex-valued) function in an open set* D *of the plane* **C**. *If f has the mean value property and if $|f|$ has a relative maximum at a point $a \in D$* (i.e. *if $|f(z)| \leqslant |f(a)|$ for any z sufficiently near to a*), *then f is constant in a neighbourhood of a.*

Proof. If $f(a) = 0$, the theorem is obvious; suppose then that $f(a) \neq 0$; by multiplying f by a complex constant if necessary, we can reduce the theorem to the case when $f(a)$ is real and > 0, which we shall assume from now on. For sufficiently small $r \geqslant 0$, let

$$M(r) = \sup_\theta |f(a + re^{i\theta})|.$$

For sufficiently small $r \geqslant 0$, we have $M(r) \leqslant f(a)$ by hypothesis. Moreover, the mean value property gives

$$f(a) = \frac{1}{2\pi}\int_0^{2\pi} f(a + re^{i\theta})\, d\theta, \tag{2.1}$$

whence $f(a) \leqslant M(r)$ and consequently $f(a) = M(r)$. It follows that the function

$$g(z) = \text{Re}\,(f(a) - f(z))$$

is $\geqslant 0$ for sufficiently small $|z - a| = r$, and that $g(z) = 0$ if and only if $f(z) = f(a)$. By (2.1), the mean value of $g(z)$ on the circle

$$|z - a| = r$$

is zero; since g is continuous and $\geqslant 0$, this requires that g is identically zero on this circle, and, consequently, $f(z) = f(a)$ when $|z - a| = r$ is sufficiently small. This completes the proof.

COROLLARY. *Let* D *be a bounded, connected, open set of the plane* **C**; *let f be a (complex-valued) continuous function defined in the closure* $\overline{D}$ *and having the mean value property in* D; *and, let* M *be the upper bound of $|f(z)|$ when z describes the frontier of* D. *Then,*

(i) $|f(z)| \leqslant M$ *for* $z \in D$;

(ii) *if* $|f(a)| = M$ *at a point* $a \in D$, *f is constant.*

Proof. Let M′ be the upper bound of $|f(z)|$ for $z \in \overline{D}$, a bound which is attained at at least one point a of the compact set $\overline{D}$ (since $|f(z)|$ is continuous). If $a \in D$, f is constant in some neighbourhood of a by theorem 1; theorem 1 also shows that the subset of D where f takes the value $f(a)$ is open, and, as it is obviously closed and non-empty, this subset must be the whole of D (because D is connected); since f is continuous in $\overline{D}$, we also have $f(z) = f(a)$ for $z \in \bar{D}$ which shows that $M = M'$ and establishes statements (i) and (ii). The other case to be proved is when $|f(a)| \neq M'$ for any point $a \in D$; but, then $M = M'$ (which proves (i)), and (ii) is trivially true because we do not have $|f(a)| = M$ for any point a of D.

Note. The maximum modulus principle is applied especially to the following case : if a continuous function f in a closed disc is holomorphic in the interior of the disc, the upper bound of $|f|$ on the boundary of the disc bounds $|f|$ above in the interior of the disc. In particular, in the Cauchy inequalities (1. 3), $M(r)$ is not only the upper bound of $|f(z)|$ for $|z| = r$ but also for $|z| \leqslant r$.

3. Schwarz' Lemma

THEOREM (Schwarz' Lemma). *Let $f(z)$ be a holomorphic function in the disc $|z| < 1$ and suppose that*

$$f(0) = 0, \qquad |f(z)| < 1 \qquad \textit{for } |z| < 1.$$

Then :

1° *we have $|f(z)| \leqslant |z|$ for $|z| < 1$;*

2° *if, for a $z_0 \neq 0$, the equality $|f(z_0)| = |z_0|$ holds, then*

$$f(z) = \lambda z \qquad \textit{identically and} \qquad |\lambda| = 1.$$

Proof. In the Taylor expansion $f(z) = \sum_{n \geqslant 0} a_n z^n$, the coefficient a_0 is is zero because $f(0) = 0$. It follows that $f(z)/z$ is holomorphic for $|z| < 1$. Since $|f(z)| < 1$ by hypothesis, we have

$$\left|\frac{f(z)}{z}\right| < \frac{1}{r} \qquad \text{for} \qquad |z| = r.$$

This inequality holds also for $|z| \leqslant r$ because of the maximum modulus principle. If we fix z in the disc $|z| < 1$, we have $|f(z)| < |z|/r$ for any $r \geqslant |z|$ and < 1. In the limit, we have then $|f(z)| \leqslant |z|$, which establishes assertion 1° of the theorem. If $|f(z_0)| = |z_0|$ for some $z_0 \neq 0$, the holomorphic function $f(z)/z$ attains the upper bound of its modulus at a point interior to the disc $|z| < 1$; thus, by the maximum modulus principle, this function is contant and we have then the identity $f(z)/z = \lambda$, $|\lambda| = 1$. This completes the proof.

4. Laurent's Expansion

1. Laurent's series

Here we consider formal power series $\sum_n a_n X^n$, where the (formal) summation is taken over all integers n, positive, negative, or 0. To such a series, we associate two formal series (in the usual sense), $\sum_{n \geqslant 0} a_n X^n$ and $\sum_{n<0} a_n X^{-n}$. Let ρ_1 and $\frac{1}{\rho_2}$ be the radii of convergence of these two series. Consider the convergent series

$$(1.1) \qquad f_1(z) = \sum_{n \geqslant 0} a_n z^n \quad \text{for} \quad |z| < \rho_1,$$

$$(1.2) \qquad f_2(z) = \sum_{n<0} a_n z^n \quad \text{for} \quad |z| > \rho_2.$$

We shall show that $f_2(z)$ is a *holomorphic* function of z. Put $z = \frac{1}{u}$; the function

$$g(u) = \sum_{n>0} a_{-n} u^n$$

is holomorphic for $|u| < 1/\rho_2$ and its derivative is given by the formula

$$g'(u) = \sum_{n>0} n a_{-n} u^{n-1}.$$

The formula for the differentiation of a composite function shows that $f_2(z)$ has a derivative equal to

$$f_2'(z) = -\frac{1}{z^2} g'(1/z) = \sum_{n<0} n a_n z^{n-1}.$$

Hence, series (1.2) is differentiable term by term for $|z| > \rho_2$. Suppose from now on that $\rho_2 < \rho_1$. Then, the sum $f(z)$ of the series

$$(1.3) \qquad \sum_{-\infty<n<+\infty} a_n z^n$$

is holomorphic in the annulus $\rho_2 < |z| < \rho_1$ and its derivative $f'(z)$ is the sum of the series $\sum n a_n z^{n-1}$ obtained by differentiating term by term.

The series $\sum a_n z^n$ is called the *Laurent series* in the annulus $\rho_2 < |z| < \rho_1$.

In the above, we do not exclude the case where $\rho_2 = 0$, nor the case where $\rho_1 = +\infty$. The convergence of series (1.3) is normal in any annulus $r_2 \leqslant |z| \leqslant r_1$, with

$$\rho_2 < r_2 < r_1 < \rho_1.$$

2. Laurent series expansion of a function holomorphic in an annulus

Definition. A function $f(z)$ defined in an annulus

$$\rho_2 < |z| < \rho_1$$

is said to have a *Laurent expansion* in this annulus if there is a Laurent series $\sum a_n z^n$ which converges in this annulus and whose sum is equal to $f(z)$ at any point of the annulus.

By the results of no. 1, $f(z)$ is then holomorphic in the annulus and the convergence is normal in any closed annulus $r_2 \leqslant |z| \leqslant r_1$ such that

$$\rho_2 < r_2 < r_1 < \rho_1;$$

moreover, we shall show that the Laurent series, if it exists, is unique. For, put $z = re^{i\theta}$ $(\rho_2 < r < \rho_1)$; by integrating the normally convergent expansion

$$f(re^{i\theta}) = \sum_{-\infty < n < +\infty} a_n r^n e^{in\theta}$$

term by term with respect to θ, we obtain, exactly as in § 1 (no. 1), the integral formula

$$(2.1) \quad a_n r^n = \frac{1}{2\pi}\int_0^{2\pi} e^{-in\theta} f(re^{i\theta})\, d\theta, \qquad \text{for } n \text{ an integer} \geqslant 0 \text{ or } < 0.$$

We see that, if the function f is given, the coefficients a_n of a Laurent expansion of f when it exists, are determined uniquely by relation (2.1). It is called the Laurent expansion of f.

Theorem. *Any holomorphic function $f(z)$ in an annulus $\rho_2 < |z| < \rho_1$ has a Laurent expansion in this annulus.*

Proof. Choose two numbers r_1 and r_2 such that

$$\rho_2 < r_2 < r_1 < \rho_1.$$

We shall show that there exists a Laurent series which converges normally in the annulus $r_2 \leqslant |z| \leqslant r_1$ and whose sum is equal to $f(z)$ in this annulus. By the uniqueness of the Laurent expansion, which follows from the integral formula (2.1), the Laurent series thus obtained will not depend on the choice of r_1 and r_2. Thus, this Laurent series will converge to $f(z)$ in the whole of the annulus $\rho_2 < |z| < \rho_1$, which will prove the theorem.

Having chosen the numbers r_1 and r_2, we choose two numbers r_1' and r_2' such that $\rho_2 < r_2' < r_2 < r_1 < r_1' < \rho_1$. Consider the compact annulus

$$r_2' \leqslant |z| \leqslant r_1'$$

whose oriented boundary is the difference of the circle γ_1 of centre o and radius r_1' described in the positive sense, and the circle γ_2 of centre o and radius r_2' described in the positive sense. By Cauchy's integral formula (chapter II, § 2, theorem 5), we have, for $r_2 \leqslant |z| \leqslant r_1$,

(2. 2) $$f(z) = \frac{1}{2\pi i}\int_{\gamma_1}\frac{f(t)\,dt}{t-z} - \frac{1}{2\pi i}\int_{\gamma_2}\frac{f(t)\,dt}{t-z}.$$

Consider the first integral; we have $|t| = r_1'$ and $|z| \leqslant r_1 < r_1'$; we can then write the series expansion

$$\frac{1}{t-z} = \sum_{n\geqslant 0}\frac{z^n}{t^{n+1}},$$

which converges normally when t describes the circle of centre o and radius r_1'. We replace $\frac{1}{t-z}$ in the first integral by this series; we can integrate it term by term because of the normal convergence, whence

(2. 3) $$\frac{1}{2\pi i}\int_{\gamma_1}\frac{f(t)\,dt}{t-z} = \sum_{n\geqslant 0} a_n z^n,$$

where

(2. 4) $$a_n = \frac{1}{2\pi i}\int_{\gamma_1}\frac{f(t)\,dt}{t^{n+1}}, \qquad n \geqslant 0.$$

Consider now the second integral; we have

$$|t| = r_2' \qquad \text{and} \qquad |z| \geqslant r_2 > r_2',$$

so

$$\frac{1}{t-z} = -\frac{1}{z}\,\frac{1}{1-t/z} = -\sum_{n<0}\frac{z^n}{t^{n+1}}.$$

Replace $\frac{1}{t-z}$ in the second integral by this series; since this series converges normally, we can integrate it term by term to obtain

(2. 5) $$-\frac{1}{2\pi i}\int_{\gamma_2}\frac{f(t)\,dt}{t-z} = \sum_{n<0} a_n z^n,$$

where

(2. 6) $$a_n = \frac{1}{2\pi i}\int_{\gamma_2}\frac{f(t)\,dt}{t^{n+1}}, \quad n < 0.$$

Finally, relation (2. 2) shows that

$$f(z) = \sum_{-\infty<n<+\infty} a_n z^n \qquad \text{for} \qquad r_2 \leqslant |z| \leqslant r_1,$$

the convergence being normal. The theorem is thus proved.

3. Decomposition of a holomorphic function in an annulus

Proposition 3. 1. *Given a holomorphic function* $f(z)$ *in an annulus* $\rho_2 < |z| < \rho_1$, *there exists a holomorphic function* $f_1(z)$ *in the disc* $|z| < \rho_1$ *and a holomorphic function* $f_2(z)$ *for* $|z| > \rho_2$ *such that*

$$f(z) = f_1(z) + f_2(z). \tag{4. 1}$$

This decomposition is unique if we stipulate that the function f_2 *tends to* 0 *as* $|z|$ *tends to* ∞.

For, let $f(z) = \sum_{-\infty < n < +\infty} a_n z^n$ be the Laurent expansion of f. Put

$$f_1(z) = \sum_{n \geqslant 0} a_n z^n, \qquad f_2(z) = \sum_{n < 0} a_n z^n. \tag{4. 2}$$

Relation (4. 1) is obviously satisfied, and $|f_2(z)|$ tends to 0 as $|z|$ tends to ∞. Suppose that

$$f(z) = g_1(z) + g_2(z)$$

is another such decomposition; let us show that $f_1 = g_1$ and $f_2 = g_2$. Let h be the holomorphic function which is equal to $f_1 - g_1$ for $|z| < \rho_1$ and equal to $g_2 - f_2$ for $|z| > \rho_2$; this function h is holomorphic in the whole plane and tends to 0 as $|z|$ tends to ∞. By the maximum modulus principle (§ 2, no. 2), or by Liouville's theorem (§ 1, no. 2), the function h is identically zero. This completes the proof.

4. Cauchy's inequalities; application to the study of isolated singularities

Consider the integral formula (2. 1). If $M(r)$ denotes the upper bound of $|f(z)|$ for $|z| = r$, the right hand side of (2. 1) has its modulus bounded above by $M(r)$, whence the Cauchy inequality

$$|a_n| \leqslant \frac{M(r)}{r^n}, \quad \text{with } n \text{ an integer } \geqslant 0 \text{ or } < 0. \tag{4. 1}$$

We shall consider a holomorphic function $f(z)$ in the punctured disc $0 < |z| < \rho$. We ask if this function can be extended to a holomorphic function in the complete disc $|z| < \rho$, centre included. This extension is obviously unique if it exists (by the principle of analytic continuation, or, in this case, simply because of continuity).

Proposition 4. 1. *A necessary and sufficient condition for this extension to be possible is that the function* $f(z)$ *is bounded in some neighbourhood of* 0.

The condition is obviously necessary. We shall show that it is sufficient.

In the punctured disc $0 < |z| < \rho$, the function f has a Laurent expansion $\sum_{-\infty<n<+\infty} a_n z^n$. By hypothesis, there exists a number $M > 0$ which bounds $|f(z)|$ above for $|z| = r$ with any sufficiently small r. By Cauchy's inequality (4. 1), we have

$$|a_n| \leqslant \frac{M}{r^n}$$

for all small r, and for $n < 0$ this implies that $a_n = 0$. Thus the Laurent expansion of f reduces to a Taylor series and this defines the required extension of $f(z)$.

Definition. Let $f(z)$ be a holomorphic function in the punctured disc $0 < |z| < \rho$. The origin 0 is said to be an *isolated singularity* of f if the function f cannot be extended to a holomorphic function on the entire disc $|z| < \rho$.

A necessary and sufficient condition for 0 to be an isolated singularity is that the coefficients a_n in the Laurent expansion are not all zero for $n < 0$. We see that there are two possible cases :

1 *st. case* : there are only a finite number of integers $n < 0$ for which $a_n \neq 0$. In this case, there is a positive integer n such that $z^n f(z)$ is a holomorphic function $g(z)$ in some neighbourhood of the origin. Thus $f(z) = g(z)/z^n$ is *meromorphic* in some neighbourhood of the origin.

2 *nd. case* : there is an infinity of integers $n < 0$ such that $a_n \neq 0$. In this case the function $f(z)$ is not a meromorphic function in a neighbourhood of the origin.

Definition. In the first case, we say that the point 0 is a *pole* of the function f; in the second case, we say that 0 is an *essential singularity* of the function f.

THEOREM (Weierstrass). *If 0 is an isolated essential singularity of a holomorphic function $f(z)$ in the punctured disc $0 < |z| < \rho$, then, for any $\varepsilon > 0$, the image of the punctured disc $0 < |z| < \varepsilon$ under f is everywhere dense in the plane* **C**.

Proof. We use *reductio ad absurdum* by supposing that there exists a disc centred at a of radius $r > 0$ which is *outside the image* of the punctured disc $0 < |z| < \varepsilon$ under f. We have then

$$|f(z) - a| \geqslant r \quad \text{for} \quad 0 < |z| < \varepsilon. \tag{4. 2}$$

The function $g(z) = \dfrac{1}{f(z) - a}$ will then be holomorphic and bounded in the punctured disc $0 < |z| < \varepsilon$. By proposition 4. 1, this function can be extended to a holomorphic function in the disc $|z| < \varepsilon$, again denoted by $g(z)$. Thus, $\frac{1}{g(z)}$ will be meromorphic in the disc $|z| < \varepsilon$ and $f(z) = a + \frac{1}{g(z)}$ will also be meromorphic, which contradicts the hypothesis that 0 is an essential singularity of $f(z)$.

Note. The case when z_0 is an essential singularity is obviously reduced to the case when $z_0 = 0$ by replacing z by $z - z_0$.

The following theorem, which we shall not prove, is much more precise than the Weierstrass theorem :

PICARD'S THEOREM. *If* 0 *is an isolated essential singularity of the holomorphic function* $f(z)$, *then the image by* f *of any punctured disc* $0 < |z| < \varepsilon$ *is either the whole plane* **C**, *or the plane* **C** *with one point missing.*

Example. The function $e^{1/z} = \sum_{n \geqslant 0} \frac{1}{n!} \frac{1}{z^n}$ is holomorphic in the punctured plane $z \neq 0$ and has an isolated essential singularity at the origin since the coefficient of $\frac{1}{z^n}$ is $\neq 0$ for all $n \geqslant 0$. This function never takes the value 0; a worthwhile exercise is to show that it takes any value $\neq 0$ in any punctured disc $0 < |z| < \varepsilon$.

5. Introduction of the Point at Infinity. Residue Theorem

1. RIEMANN SPHERE

In the space $\mathbf{R}^3$, let x, y, u be the coordinates of a point and let us consider the unit sphere $\mathbf{S}_2$,

$$x^2 + y^2 + u^2 = 1.$$

The sphere $\mathbf{S}_2$, with the topology induced by that of the space $\mathbf{R}^3$, is a compact space since $\mathbf{S}_2$ is a bounded closed subset of $\mathbf{R}^3$. Let P and P′ be two points of $\mathbf{S}_2$ whose coordinates are respectively (0, 0, 1) and (0, 0, −1). We shall consider stereographic projection from the pole P. It associates with any point M of $\mathbf{S}_2$ other than P the point of the plane $u = 0$ collinear with P and M. The complex coordinate z of this point is given by the formula

$$z = \frac{x + iy}{1 - u}, \qquad (1.1)$$

where x, y, u are the coordinates of the point M. Similary, we consider stereographic projection from the pole P′ but we take the point of the plane $u = 0$ which is the complex conjugate of the point corresponding to M (x, y, u) under this stereographic projection. Its complex coordinate z' is given by the formula

$$z' = \frac{x - iy}{1 + u}. \tag{1. 2}$$

Note that, for any point M (x, y, u) other than P or P′, the corresponding complex numbers z and z' are related by

$$zz' = 1. \tag{1. 3}$$

The mapping $(x, y, u) \to z$ is a homeomorphism of $\mathbf{S}_2$ — P onto $\mathbf{C}$; we say that we have a chart of $\mathbf{S}_2$ — P on the complex plane $\mathbf{C}$. Similarly, the mapping $(x, y, u) \to z'$ is a chart of $\mathbf{S}_2$ — P′ onto the complex plane $\mathbf{C}$. Provided with these two charts, $\mathbf{S}_2$ is called the *Riemann sphere.*

Let D be an open set of $\mathbf{S}_2$. We say that a function f defined in D is *holomorphic* in D if, in some neighbourhood of any point M $\in$ D distinct from P, it can be expressed as a holomorphic function of z, and if, in some neighbourhood of any point M $\in$ D distinct from P′, it can be expressed as a holomorphic function of z'. We note that, in a neighbourhood of a point distinct from both P and P′, any holomorphic function of z is a holomorphic function of z', and conversely, because of relation (1. 3). By means of relation (1. 1), we shall always identify the complex plane $\mathbf{C}$ with the sphere $\mathbf{S}_2$ with the point P excluded. We see that $\mathbf{S}_2$ is obtained by adding 'a point at infinity', to $\mathbf{C}$. To study a function in a neighbourhood of the point at infinity P we use the complex variable $z' = 1/z$, which is zero at the point P. The open sets $|z| > r$ in $\mathbf{C}$ form a fundamental system of neighbourhoods of the point at infinity. A function $f(z)$ defined in such an open set is 'holomorphic at infinity' if, by the change of variable $z = 1/z'$, it is expressed as a holomorphic function of z' for $|z'| < 1/r$.

Similarly, a function $f(z)$ is *meromorphic at infinity* if, when expressed as a function of z', it is meromorphic in a neighbourhood of $z' = 0$. Finally, a holomorphic function $f(z)$ for $|z| > r$ has an isolated essential singularity at the point at infinity if the function $f(1/z')$ has an isolated essential singularity at the origin $z' = 0$.

If

$$f(z) = \sum_n a_n z^n$$

is the Laurent expansion of $f(z)$ for $|z| > r$, a necessary and sufficient condition for the point at infinity to be a pole of f is that $a_n = 0$ for all the integers $n \geqslant 0$ except for a finite number of them; the condition for an essential singularity at the point at infinity is that there exist an infinity of integers $n \geqslant 0$ such that $a_n \neq 0$.

The concepts of *differentiable path*, *closed path*, and *oriented boundary of a compact set* can be defined for the sphere $\mathbf{S}_2$.

2. Residue theorem.

First, let us consider a holomorphic function $f(z)$ in an annulus $\rho_2 < |z| < \rho_1$ centred at the origin.

PROPOSITION 2. 1. *If γ is a closed path contained in the annulus, then*

$$\frac{1}{2\pi i}\int_\gamma f(z)\,dz = \mathrm{I}(\gamma, 0)\,a_{-1}, \tag{2. 1}$$

where $\mathrm{I}(\gamma, 0)$ is the index of the path γ with respect to the origin 0 and a_{-1} is the coefficient of $1/z$ in the Laurent expansion of f.

Proof. We have

$$f(z) = a_{-1}/z + g(z),$$

where

$$g(z) = \sum_{n\neq -1} a_n z^n$$

is holomorphic on the annulus and has a primitive in it equal to

$$\sum_{n\neq -1} \frac{a_n}{n+1} z^{n+1} \qquad \text{(cf. § 4, n° 1).}$$

Thus, we have the relation

$$\int_\gamma f(z)\,dz = a_{-1}\int_\gamma dz/z + \int_\gamma g(z)\,dz. \tag{2. 2}$$

But, $\int_\gamma g(z)\,dz = 0$ since g has a primitive, and

$\int_\gamma dz/z = 2\pi i\,\mathrm{I}(\gamma, 0)$ by the definition of the index.

These two relations, along with (2. 2), give (2. 1).

Formula (2. 1) is applied particularly in the case when the function f has an isolated singularity at the origin 0 (either a pole, or an essential singularity). In this case, γ is a closed path in some neighbourhood of 0 which does not pass through 0. The coefficient a_{-1} of the Laurent expansion is then called the *residue* of the function f at the singular point 0. In particular, if γ is a circle centred at 0 with small radius described in the positive sense then

$$\int_\gamma f(z)\,dz = 2\pi i a_{-1}. \tag{2. 3}$$

The residue at any isolated singularity situated at any point of the plane $\mathbf{C}$ is defined in a similar way.

A *residue at the point at infinity* needs a special definition : let $f(z)$ be a holomorphic function for $|z| > r$ and put $z = \frac{1}{z'}$; then,

$$f(z)\, dz = -\frac{1}{z'^2} f\left(\frac{1}{z'}\right) dz'.$$

By definition, the residue of f at the point at infinity is equal to the residue of the function $-\frac{1}{z'^2} f\left(\frac{1}{z'}\right)$ at the point $z' = 0$. Thus, if $\sum_n a_n z^n$ is the Laurent expansion of $f(z)$ in a neighbourhood of the point at infinity, the residue of f at infinity is $-a_{-1}$.

RESIDUE THEOREM. *Let D be an open set of the Riemann sphere* $\mathbf{S}_2$ *and let f be a holomorphic function in D except perhaps at isolated points which are singularities of f. Let* Γ *be the oriented boundary of a compact subset A of D and suppose that* Γ *does not pass through any singularities of f, or the point at infinity. Then, only a finite number of singularities* z_k *are contained in A, and*

$$\int_\Gamma f(z)\, dz = 2\pi i\left(\sum_k \operatorname{Res}(f, z_k)\right), \tag{2.4}$$

where $\operatorname{Res}(f, z_k)$ *denotes the residue of the function f at the point* z_k; *the summation extends over all the singularities* $z_k \in A$ *including the point at infinity if it qualifies.*

Proof. We distinguish between the two cases where the point at infinity belongs or does not belong to A.

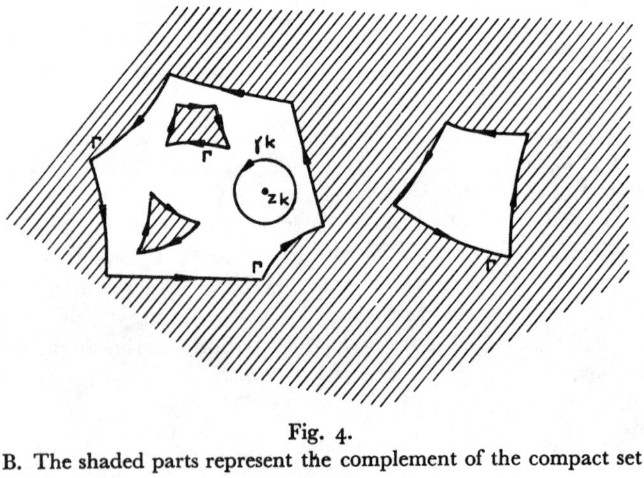

Fig. 4.

N. B. The shaded parts represent the complement of the compact set A.

I st. case. The point at infinity does not belong to A; A is then a (bounded) compact set of the plane $\mathbf{C}$ (cf. fig. 4); each singular point z_k is the centre of a closed disc S_k in the interior of A and we can choose the radii of these discs small enough for the discs to be disjoint. Let γ_k be the boundary of the disc δ_k described in the positive sense. Let A′ be the compact set obtained by removing the interiors of the above discs from A; the oriented boundary of A′ is the difference between Γ (the oriented boundary of A) and the circles γ_k. Since f is holomorphic in some neighbourhood of A′, we have (cf. chapter II, § 2, no. 8, theorem 5)

$$\int_\Gamma f(z)\, dz = \sum_k \int_{\gamma_k} f(z)\, dz. \tag{2. 5}$$

On the other hand, by (2. 3)

$$\int_{\gamma_k} f(z)\, dz = 2\pi i \operatorname{Res}(f, z_k),$$

and substituting this in (2. 5) gives the required relation (2. 4).

2 nd. case. The point at infinity belongs to A. Let $|z| \geqslant r$ be a neighbourhood of the point at infinity which does not intersect Γ and such that $f(z)$ is holomorphic in this neighbourhood (the point at infinity being excluded). Let A″ be the compact set obtained by removing the open set $|z| > r$ from A (cf. fig. 5)

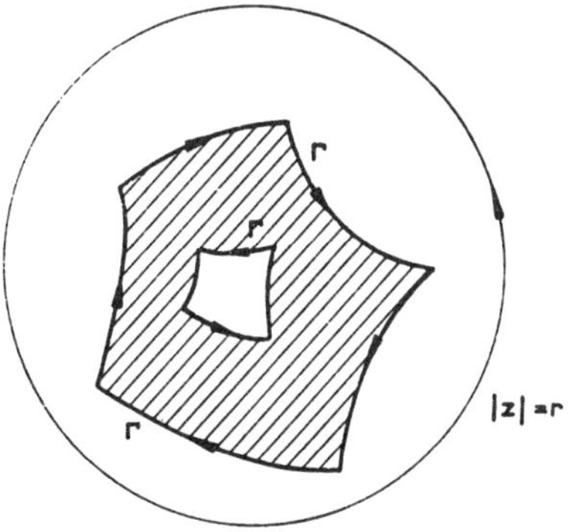

Fig. 5.
N. B. The shaded portion represents the complement of A.

The oriented boundary of A″ is the sum of the oriented boundary Γ of A and of the circle $|z| = r$ described in the positive sense. By applying the results we have proved in the first case to A″, we obtain

$$\int_\Gamma f(z)\, dz + \int_{|z|=r} f(z)\, dz = 2\pi i \sum_k \operatorname{Res}(f, z_k), \tag{2. 6}$$

where the sum on the right hand side extends over all singularities z_k

contained in A *other than the point at infinity.* Moreover, by the definition of the residue at infinity, we have

$$\int_{|z|=r} f(z)\, dz = -\,2\pi i \operatorname{Res}(f,\ \infty),$$

and this substituted in (2. 6) gives

$$\int_{\Gamma} f(z)\, dz = 2\pi i \left(\operatorname{Res}(f, \infty) + \sum_k \operatorname{Res}(f, z_k)\right)$$

which is none other than the required relation (2. 4) when the point at infinity is one of the singularities z_k.

Note. Consider in particular the case where the compact set is the whole sphere $\mathbf{S}_2$. In this case, the boundary is empty, and relation (2. 4) becomes:

(2. 7) $$\sum_k \operatorname{Res}(f,\ z_k) = 0.$$

For example, *the sum of the residues of a rational function (including the residue at infinity) is zero.*

3. Practical calculation of residues

The case of a simple pole which is not at infinity. Let z_0 be a simple pole of f; then

$$f(z) = \frac{1}{z - z_0}\, g(z),$$

where g is holomorphic in some neighbourhood of z_0 with $g(z_0) \neq 0$. Let

$$g(z) = \sum_{n \geq 0} a_n (z - z_0)^n$$

be the Taylor expansion of $g(z)$ in a neighbourhood of z_0; we see that, in the Laurent expansion of $f(z)$, the coefficient of $\dfrac{1}{z - z_0}$ is equal to $g(z_0)$. Thus,

(3. 1) $$\operatorname{Res}(f,\ z_0) = \lim_{\substack{z \to z_0 \\ z \neq z_0}} (z - z_0) f(z).$$

If f is given in the form of a quotient P/Q, where P and Q are holomorphic in a neighbourhood of z_0 and where z_0 is a simple zero of Q with $P(z_0) \neq 0$, then

(3. 2) $$\operatorname{Res}(f,\ z_0) = \frac{P(z_0)}{Q'(z_0)},$$

Q′ denoting the derivative of Q.

Example. Let $f(z) = \frac{e^{iz}}{z^2 + 1}$; the function has two simple poles $z = \pm i$; we have $P/Q' = \frac{1}{2z} e^{iz}$, and, consequently, the residue of f at the pole i is equal to $-\frac{i}{2e}$.

The case of a multiple pole. Let $f(z) = \frac{1}{(z - z_0)^k} g(z)$, where $g(z)$ is holomorphic in a neighbourhood of the point z_0 with $g(z_0) \neq 0$. The residue of $f(z)$ is equal to the coefficient of $(z - z_0)^{k-1}$ in the Taylor expansion of $g(z)$ at the point z_0. The problem is reduced, then, to finding a limited expansion of $g(z)$. To this end, it is often convenient to take a new variable $t = z - z_0$.

Example. Let $f(z) = \frac{e^{iz}}{z(z^2 + 1)^2}$ and let us calculate residue of $f(z)$ at the double pole $z = i$. In this case,

$$g(z) = \frac{e^{iz}}{z(z + i)^2}.$$

Put $z = i + t$, so we must find coefficient of t in the Taylor expansion of

$$h(t) = \frac{e^{i(i+t)}}{(i + t)(2i + t)^2}.$$

It is sufficient to write down the limited expansion of degree 1 of each of the terms

$$e^{i(i+t)} = e^{-1}(1 + it + \cdots),$$
$$(i + t)^{-1} = -i(1 - it)^{-1} = -i(1 + it + \cdots),$$
$$(2i + t)^{-2} = -\frac{1}{4}\left(1 - \frac{i}{2}t\right)^{-2} = -\frac{1}{4}(1 + it + \cdots),$$

whence

$$h(t) = \frac{i}{4e}(1 + 3it + \cdots),$$

and the required residue is $-\frac{3}{4e}$.

Application. Residue of a logarithmic derivative. Let $f(z)$ be a meromorphic function in a neighbourhood of z_0. We propose to find the residue of the logarithmic derivative f'/f at the point z_0. We have

$$f(z) = (z - z_0)^k g(z)$$

where g is holomorphic at the point z_0 and $g(z_0) \neq 0$; the integer k is $\geqslant 0$

if f is holomorphic at z_0, and $k < 0$ if z_0 is a pole of f; taking the logarithmic derivative of the two sides gives

$$f'/f = \frac{k}{z - z_0} + g'/g;$$

thus f'/f has z_0 as a *simple pole* and *the residue of this pole is equal to the integer k*, the order of multiplicity of the zero or pole z_0 (counted positively for a zero and negatively for a pole).

4. Application to finding the number of poles and zeros of a meromorphic function.

Proposition 4. 1. *Let $f(z)$ be a meromorphic function which is not constant in an open set D and let Γ be the oriented boundary of a compact set K contained in D. Suppose that the function f has no poles on Γ and does not take the value a on Γ. Then,*

$$\frac{1}{2\pi i}\int_{\Gamma}\frac{f'(z)\,dz}{f(z)-a} = \mathrm{Z} - \mathrm{P}, \tag{4. 1}$$

where Z denotes the sum of the orders of multiplicity of the roots of the equation

$$f(z) - a = 0$$

contained in K, *and* P *denotes the sum of the orders of multiplicity of the poles of f contained in* K.

This is an immediate consequence of the residue theorem and of the explicit calculation of the residues of the function $\dfrac{f'(z)}{f(z)-a}$.
In particular, when f is holomorphic, the integral on the left hand side of (4. 1) is equal to the number of zeros of $f(z) - a$ contained in K, it being understood that each zero is counted as many times as its order of multiplicity.
You will notice that the value of the integral on the left hand side of (4. 1) is equal to the quotient by 2π of *the variation of the argument of* $f(z) - a$ when z describes the closed path Γ (cf. chapter II, § 1, no. 5).

Proposition 4. 2. *Let z_0 be a root of order k of the equation $f(z) = a$, f being a non-constant, holomorphic function in some neighbourhood of z_0. For any sufficiently small neighbourhood V of z_0, and for any b sufficiently near to a and $\neq a$, the equation $f(z) = b$ has exactly k simple solutions in* V.

For, let γ be a circle centred at z_0 with sufficiently small radius to ensure that z_0 is the only solution to the equation $f(z) = a$ contained in the closed disc bounded by γ. Suppose also that the radius of γ is sufficiently small to ensure that $f'(z)$ is $\neq 0$ at any point of the disc except the centre z_0. We consider the integral

$$\frac{1}{2\pi i}\int_\gamma \frac{f'(z)\,dz}{f(z)-b}. \tag{4. 2}$$

We know that (4. 2) remains constant when b varies in a connected component of the complement of the image of γ under f (cf. chapter II, § 1, no. 8). Thus, for any b sufficiently near to a, we have

$$\frac{1}{2\pi i}\int_\gamma \frac{f'(z)\,dz}{f(z)-b} = \frac{1}{2\pi i}\int_\gamma \frac{f'(z)\,dz}{f(z)-a} = k,$$

and, consequently, the equation $f(z) = b$ has exactly k roots in the interior of γ, if each root is counted with its order of multiplicity. But, for b sufficiently near to a but $\neq a$, the roots of the equation $f(z) = b$ are all simple because the derivative $f'(z)$ is $\neq 0$ at any point of z sufficiently near to z_0 and $\neq z_0$. Hence, proposition 4. 2 is proved.

5. Application to doubly periodic functions

Let e_1 and e_2 be two complex numbers, which are linearly independent over the real field **R**, that is to say, such that $e_1 \neq 0$ and that their quotient e_2/e_1 is not real. The totality of vectors of the form $n_1e_1 + n_2e_2$, where n_1 and n_2 are arbitrary integers, forms a discrete subgroup Ω of the additive group of the field **C**. We say that a function $f(z)$ defined on the plane has the group Ω as group of periods if

$$f(z + n_1e_1 + n_2e_2) = f(z) \tag{5. 1}$$

for all z and for all integers n_1 and n_2. A necessary and sufficient condition for this is that

$$f(z + e_1) = f(z) \qquad f(z + e_2) = f(z). \tag{5. 2}$$

Let z_0 be any complex number. We consider the (closed) parallelogram with vertices z_0, $z_0 + e_1$, $z_0 + e_2$, $z_0 + e_1 + e_2$. It consists of all points of the form $z_0 + t_1e_1 + t_2e_2$, where $0 \leqslant t_1 \leqslant 1$ and $0 \leqslant t_2 \leqslant 1$. Such a parallelogramm is called a *parallelogram of periods* with first vertex z_0. Let $f(z)$ now be a meromorphic function in the whole plane which has Ω as its group of periods, and choose z_0 in such a way that $f(z)$ has no poles on the boundary γ of parallelogram of periods with z_0 as first vertex. We can consider the integral $\int_\gamma f(z)\,dz$, whose value is zero because of the periodicity; for

$$\int_\gamma f(z)\,dz = \int_0^1 [f(z_0 + te_1) - f(z_0 + e_2 + te_1)]\,dt + \int_0^1 [f(z_0 + e_1 + te_2) - f(z_0 + te_2)]\,dt.$$

By appling this result to the logarithmic derivative f'/f and using proposition 4. 1, we obtain :

PROPOSITION 5. 1. *If $f(z)$ is a non-constant meromorphic function in the whole plane which has Ω as group of periods, the number of zeros of this function contained in a parallelogramm of periods is equal to the number of poles contained in the same parallelogram, if no zeros or poles of the function f occur on the boundary of the parallelogram.*

COROLLARY. *A holomorphic function in* **C** *having Ω as group of periods is constant.*

Otherwise, the number of zeros of $f(z) - a$ would be equal to the number of poles, that is zero; but, this is true for all a, which is absurd.
Moreover, consider the function $zf'(z)/(f(z) - a)$. It is not periodic, so we can no longer say that its integral round the boundary γ of some parallelogram of periods is zero. We shall show that the value of the integral

$$\frac{1}{2\pi i}\int_{\gamma}\frac{zf'(z)\,dz}{f(z)-a} \tag{5. 3}$$

belongs to the group Ω of periods. For, it is equal to

$$-\frac{e_2}{2\pi i}\int_{\gamma_1}\frac{f'(z)\,dz}{f(z)-a}+\frac{e_1}{2\pi i}\int_{\gamma_2}\frac{f'(z)\,dz}{f(z)-a},$$

where γ_1 denotes the side of the parallelogram starting at z_0 and ending at $z_0 + e_1$, and γ_2 denotes the side of the parallelogram starting at z_0 and ending at $z_0 + e_2$. However, the values of the integrals $\frac{1}{2\pi i}\int_{\gamma_1}\frac{f'(z)\,dz}{f(z)-a}$ and $\frac{1}{2\pi i}\int_{\gamma_2}\frac{f'(z)\,dz}{f(z)-a}$ are *integers*.

On the other hand, the integral (5. 3) is equal to the sum of the residues of the function $zf'(z)/(f(z) - a)$. Let us calculate these residues. The poles are at most the poles of $f(z)$ and the zeros of $f(z) - a$. If β_i is a pole and k is its order of multiplicity, then the residue for this pole is equal to $-k\beta_i$. Similarly, the residue of a zero α_i of multiplicity k of $f(z) - a$ is equal to $k\alpha_i$.

This is summed up by the following :

PROPOSITION 5. 2. *Let $f(z)$ be a non-constant, meromorphic function in the whole plane having Ω as group of periods. For any complex number a, we have*

$$\sum_i \alpha_i \equiv \sum_i \beta_i \qquad \text{mod. } \Omega,$$

where the α_i denote the roots of the equation $f(z) = a$ (each occurring as often *as its multiplicity) and the* β_i *denote the poles (occurring as often as their multiplicity) contained in a parallelogram of periods.*

In particular, the sum $\sum_i \alpha_i$ taken modulo Ω is independent of a.

6. Evaluation of Integrals by the Method of Residues

We propose to evaluate definite integrals without finding a primitive of the integrand, but by interpreting the value of the integral as the sum of the residues at the singular points of a suitably chosen holomorphic function. There is no general method of dealing with this problem. We shall limit ourselves to some classical types and indicate, for each of them, the procedure by which the problem can be transformed into a residue calculation.

1 *st type.* Consider an integral of the form

$$I = \int_0^{2\pi} R(\sin t, \cos t)\, dt,$$

where $R(x, y)$ is a rational function without a pole on the circle $x^2 + y^2 = 1$. Put $z = e^{it}$, so that z describes the unit circle as t increases form 0 to 2π. Thus, I is equal to $2\pi i$ times the sum of the residues of the function

$$\frac{1}{iz} R\left(\frac{1}{2i}\left(z - \frac{1}{z}\right), \frac{1}{2}\left(z + \frac{1}{z}\right)\right)$$

at the poles contained in the unit disc.

We then have

$$I = 2\pi \sum \operatorname{Res}\left\{\frac{1}{z} R\left(\frac{1}{2i}\left(z - \frac{1}{z}\right), \frac{1}{2}\left(z + \frac{1}{z}\right)\right)\right\},$$

the sum extending over poles contained in the unit disc.

Example. Let $\int_0^{2\pi} \frac{dt}{a + \sin t}$, where a is a real number > 1. Then,

$$I = 2\pi \sum \operatorname{Res} \frac{2i}{z^2 + 2iaz - 1}.$$

The only pole contained in the unit disc is $z_0 = -ia + i\sqrt{a^2 - 1}$; its residue is $\frac{i}{z_0 + ia} = \frac{1}{\sqrt{a^2 - 1}}$, so $I = \frac{2\pi}{\sqrt{a^2 - 1}}$.

2 *nd type.* Consider an integral of the form

$$I = \int_{-\infty}^{+\infty} R(x)\, dx,$$

where R is a rational function without a real pole. We also need to assume that the integral is convergent, and a necessary and sufficient condition for this is that the principal part of $R(x)$ at infinity is of the form $\frac{1}{x^n}$ with the integer $n \geqslant 2$. An equivalent condition is that

(2. 1) $$\lim_{|x| \to \infty} xR(x) = 0.$$

To calculate this integral I, we shall integrate the function $R(z)$ of the complex variable z along the boundary γ of a half-disc of centre o and radius r situated in the half-plane $y \geqslant 0$ (fig. 6). For sufficiently large r,

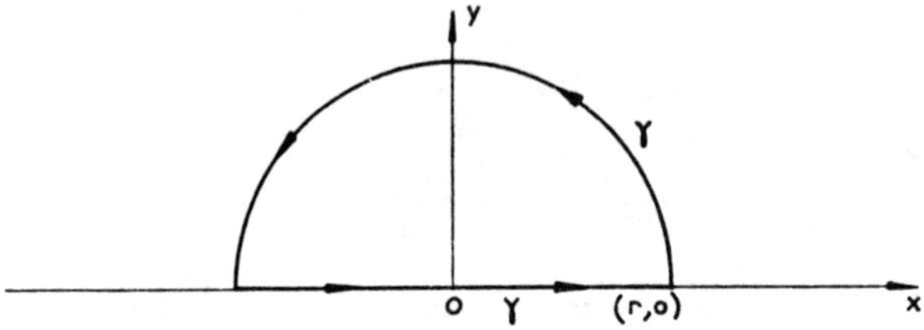

Fig. 6.

the function $R(z)$ is holomorphic on the boundary γ and the integral $\int_\gamma R(z)\, dz$ is equal to the sum of the residues of the poles of R which are inside γ. Then

(2. 2) $$\int_{-r}^{+r} R(x)\, dx + \int_{\delta(r)} R(z)\, dz = 2\pi i \sum \text{Res}\,(R(z)),$$

where $\delta(r)$ denotes the half-circle centred at o of radius r described in the positive sense, and where the summation extends over the residues of poles situated in the half-plane $y > 0$. As r tends to $+\infty$, the first integral on the left hand side of (2. 2) tends to I; we shall show that the second integral on the left hand side of (2. 2) tends to o. This will give

(2. 3) $$\int_{-\infty}^{+\infty} R(x)\, dx = 2\pi i \sum \text{Res}\,(R(z)),$$

the sum extending over all the poles of R in the upper half-plane $y > 0$. Similarly,

$$\int_{-\infty}^{+\infty} R(x)\, dx = -\, 2\pi i \sum \text{Res}\,(R(z)),$$

the sum this time being taken over all the poles in the lower half-plane $y < 0$.

It remains to be proved that $\int_{\delta(r)} R(z)\,dz$ tends to 0 as r tends to $+\infty$, which will be an immediate consequence of the following lemma:

LEMMA 1. *Let $f(z)$ be a continuous function defined in the sector*

$$\theta_1 \leqslant \theta \leqslant \theta_2,$$

r and θ denoting the modulus and argument of z. If

$$\lim_{|z|\to\infty} z f(z) = 0 \quad (\theta_1 \leqslant \arg z \leqslant \theta_2),$$

then the integral $\int f(z)\,dz$ extended over the arc of the circle $|z| = r$ contained in the sector tends to 0 as r tends to $+\infty$.

For, let $M(r)$ be the upper bound of $|f(z)|$ on the arc of the circle $|z| = r$. Then

$$\left|\int f(z)\,dz\right| \leqslant M(r)\,r(\theta_2 - \theta_1),$$

and the lemma follows immediately from this.

We could prove the following lemma similarly:

LEMMA 2. *Let $f(z)$ be a continuous function defined in a sector*

$$\theta_1 \leqslant \theta \leqslant \theta_2,$$

r and θ being the modulus and argument of z. If

$$\lim_{z\to 0} z f(z) = 0 \quad (\theta_1 \leqslant \arg z \leqslant \theta_2),$$

then the integral $\int f(z)\,dz$ over the arc of the circle $|z| = r$ contained in the sector tends to 0 as r tends to 0.

Example. To evaluate the integral

$$I = \int_0^{+\infty} \frac{dx}{1 + x^6}.$$

The function $\frac{1}{1 + z^6}$ has six poles, all on the unit circle; the three poles which are in the upper half-plane are

$$e^{i\frac{\pi}{6}}, \quad e^{i\frac{\pi}{2}}, \quad e^{5i\frac{\pi}{6}}.$$

The residue of such a pole is equal to $\frac{1}{6z^5} = -\frac{z}{6}$. Hence,

$$I = \frac{1}{2}\int_{-\infty}^{+\infty} \frac{dx}{1 + x^6} = -\frac{\pi i}{6}\left(e^{i\frac{\pi}{6}} + e^{i\frac{\pi}{2}} + e^{5i\frac{\pi}{6}}\right)$$
$$= \frac{\pi}{6}\left(2 \sin\frac{\pi}{6} + 1\right) = \frac{\pi}{3}.$$

3 *rd type.* We shall study integrals of the the form

$$I = \int_{-\infty}^{+\infty} f(x)\, e^{ix}\, dx,$$

where $f(z)$ is a holomorphic function in a neighbourhood of each point of the closed half-plane $y \geqslant 0$, except perhaps at a *finite* number of points. We shall first consider the case when the singularities are not on the real axis. Then, the integral

$$\int_{-r}^{+r} f(x)\, e^{ix}\, dx$$

has a meaning, and, as r tends to $+\infty$, its value tends to

$$\int_{-\infty}^{+\infty} f(x)\, e^{ix}\, dx$$

if the latter integral is convergent.

We shall prove the following result :

PROPOSITION 3. 1. *If* $\lim\limits_{|z|\to\infty} f(z) = 0$ *for* $y \geqslant 0$, *then*

$$\lim_{r\to+\infty}\int_{-r}^{+r} f(x)\, e^{ix}\, dx = 2\pi i \sum \operatorname{Res}\left(f(z)\, e^{iz}\right), \tag{3. 1}$$

the summation extending over the singularities of $f(z)$ *contained in the upper half-plane* $y > 0$.

First, we note that, if the integral $\int_{-\infty}^{+\infty} |f(x)|\, dx$ is convergent, the proposed integral $\int_{-\infty}^{+\infty} f(x) e^{ix}\, dx$ is absolutely convergent; relation (3. 1) then gives

$$\int_{-\infty}^{+\infty} f(x) e^{ix}\, dx = 2\pi i \sum \operatorname{Res}\left(f(z)\, e^{iz}\right). \tag{3. 2}$$

The integral $\int_{-\infty}^{+\infty} f(x)\, e^{ix}\, dx$ can also be convergent without being absolutely convergent; for example it is well-known that, if the function $f(x)$ is real and monotonic for $x > 0$ and tends to 0 as x tends to $+\infty$, then the integral $\int_0^{+\infty} f(x)\, e^{ix}\, dx$ is convergent (by applying the second mean value theorem); in such a case, relation (3. 2) is again true.

Before starting the proof of proposition 3. 1, we note that $|e^{iz}| \leqslant 1$ in the half-plane $y \geqslant 0$. This leads us to integrate on the half-plane $y \geqslant 0$ along the contour already used above for the second type of integral. With the same notations as in (2. 2), we shall show that *the integral* $\int_{\delta(r)} f(z) e^{iz}\, dz$ *tends to* 0 *as r tends to* $+\infty$. Proposition 3. 1 will follow obviously from this.

If we knew that $\lim_{|z|\to\infty} z f(z) = 0$, it would be sufficient to apply lemma 1. Relation (3. 1) is thus proved in this case. For example, consider the integral

$$\int_0^{+\infty} \frac{\cos x}{x^2+1}\, dx = \frac{1}{2} \operatorname{Re}\left(\int_{-\infty}^{+\infty} \frac{e^{ix}}{x^2+1}\, dx\right);$$

its value is equal to $\pi i \sum \operatorname{Res}\left(\frac{e^{iz}}{z^2+1}\right)$, the summation extending over poles situated in the upper half-plane. There is only one pole $z = i$, it is simple, and its residue is

$$\frac{e^{-1}}{2i}, \qquad \text{whence} \qquad \int_0^{+\infty} \frac{\cos x}{x^2+1}\, dx = \frac{\pi}{2e}.$$

To prove that $\int_{\delta(r)} f(z)\, e^{iz}\, dz$ tends to zero *with only the hypothesis of proposition* 3. 1, we use the following lemma :

Lemma 3. *Let $f(z)$ be a function defined in a sector of the half-plane $y \geqslant 0$. If $\lim_{|z|\to\infty} f(z) = 0$, the integral $\int f(z)\, e^{iz}\, dz$ extended over the arc of the circle $|z| = r$ contained in the sector tends to* 0 *as r tends to* $+\infty$.

For, let us put $z = re^{i\theta}$ and let $M(r)$ be the upper bound of $|f(re^{i\theta})|$ as θ varies, the point $e^{i\theta}$ remaining in the sector. Then

$$\left|\int f(z)\, e^{iz}\, dz\right| \leqslant M(r) \int_0^{\pi} e^{-r\sin\theta}\, r\, d\theta \tag{3. 3}$$

We shall show that $\int_0^{\pi} e^{-r\sin\theta} r\, d\theta$ is bounded above by a fixed number independent of r, which will complete the proof of lemma 3. In fact,

$$\int_0^{\pi} e^{-r\sin\theta} r\, d\theta = 2\int_0^{\frac{\pi}{2}} e^{-r\sin\theta} r\, d\theta \leqslant \pi. \tag{3. 4}$$

Proof of (3. 4) : we have

$$\frac{2}{\pi} \leqslant \frac{\sin\theta}{\theta} \leqslant 1 \qquad \text{for} \qquad 0 \leqslant \theta \leqslant \frac{\pi}{2},$$

whence

$$\int_0^{\frac{\pi}{2}} e^{-r\sin\theta} r\,d\theta \leqslant \int_0^{\frac{\pi}{2}} e^{-\frac{2}{\pi}r\theta} r\,d\theta \leqslant \int_0^{+\infty} e^{-\frac{2}{\pi}r\theta} r\,d\theta = \frac{\pi}{2}.$$

Hence the proposition 3. 1 is completely proved.

We now examine the case when $f(z)$ can have singularities on the real axis. We shall limit ourselves to one example, the case when $f(z)$ has a simple pole at the origin. In this case, it is appropriate to modify the path of integration to make it by-pass the origin along a semicircle $\gamma(\varepsilon)$ of small radius $\varepsilon > 0$ centred *at the origin and situated in the upper half-plane* (fig. 7). We use the following lemma :

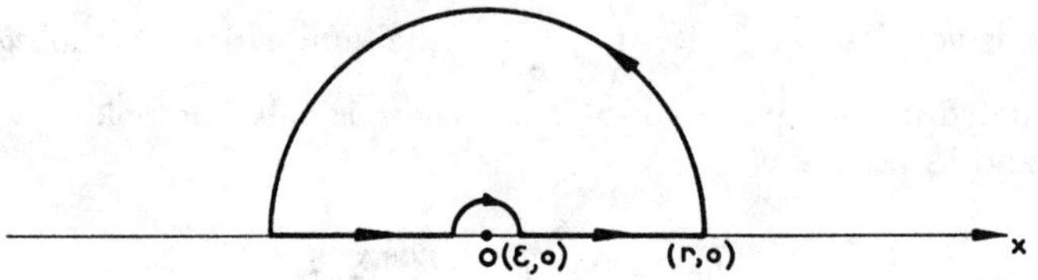

Fig. 7.

LEMMA 4. *If $z = 0$ is a simple pole of $g(z)$, then*

$$\text{(3. 4)} \qquad \lim_{\varepsilon \to 0} \int_{\gamma(\varepsilon)} g(z)\,dz = \pi i \operatorname{Res}(g, 0),$$

$\gamma(\varepsilon)$ being described in the direction of increasing argument.

For, we have $g(z) = \dfrac{a}{z} + h(z)$, where h is a holomorphic function at the origin. The integral $\int_{\gamma(\varepsilon)} h(z)\,dz$ tends to 0 as ε tends to 0, and the integral $\int_{\gamma(\varepsilon)} \dfrac{a}{z}\,dz$ is equal to $\pi i a$. This gives relation (3. 4).

This lemma will be applied to the function $g(z) = f(z)e^{iz}$.

Example. To evaluate the integral

$$I = \int_0^{+\infty} \frac{\sin x}{x}\,dx = \frac{1}{2}\int_{-\infty}^{+\infty} \frac{\sin x}{x}\,dx = \frac{1}{2i}\lim_{\varepsilon\to 0}\left[\int_{-\infty}^{-\varepsilon} \frac{e^{ix}}{x}\,dx + \int_{+\varepsilon}^{+\infty} \frac{e^{ix}}{x}\,dx\right].$$

By figure 7, this is equal to

$$\frac{1}{2i}\lim_{\varepsilon\to 0}\int_{\gamma(\varepsilon)} \frac{e^{iz}}{z}\,dz = \frac{\pi}{2}\operatorname{Res}\left(\frac{e^{iz}}{z}, 0\right) = \frac{\pi}{2}.$$

Important note. If, instead of $\int_{-\infty}^{+\infty} f(x)e^{ix}\,dx$, we had to calculate the integral $\int_{-\infty}^{+\infty} f(x)\,e^{-ix}\,dx$, then it would be necessary to integrate *in the lower half-plane* instead of the upper half-plane; for, the function $|e^{-iz}|$ is bounded in the lower half-plane $y \leqslant 0$ and it is in this half-plane that lemma 3 is applicable (*mutatis mutandis*). More generally, an integral of the form $\int_{-\infty}^{+\infty} f(x)e^{ax}\,dx$ (where a is complex constant) can be evaluated by integrating in the half-plane where $|e^{az}| \leqslant 1$.

Remember that $\sin z$ and $\cos z$ *are not bounded in any half-plane.* To evaluate integrals of the forms

$$\int_{-\infty}^{+\infty} f(x)\sin^n x\,dx, \qquad \int_{-\infty}^{+\infty} f(x)\cos^n x\,dx,$$

one always expresses the trigonometric functions in terms of complex exponentials so that the preceding methods can be applied.

4 *th type.* Consider integrals of the form

$$I = \int_0^{+\infty} \frac{R(x)}{x^\alpha}\,dx,$$

where α denotes a real number such that $0 < \alpha < 1$, and $R(x)$ is a rational function with no pole on the positive real axis $x \geqslant 0$. It is clear that such an integral converges for the lower limit of integration 0. A necessary and sufficient condition for it to converge at the upper limit $+\infty$ is that the principal part of $R(x)$ at infinity is of the form $\frac{1}{x^n}$ with $n \geqslant 1$: in other words, it is necessary and sufficient that

$$\lim_{x\to+\infty} R(x) = 0. \tag{4.1}$$

To calculate such an integral, we consider the function $f(z) = \frac{R(z)}{z^\alpha}$ of the complex variable z, defined in the plane with the positive real axis $x \geqslant 0$ excluded. Let D be the open set thus defined. It is necessary to specify the branch of z^α chosen in D: we shall take the branch of the argument of z between 0 and 2π. With this convention, integrate $\frac{R(z)}{z^\alpha}$ along the closed path $\delta(r, \varepsilon)$ defined as follows: we describe, first, the real axis from $\varepsilon > 0$ to $r > 0$, then the circle $\gamma(r)$ of centre 0 and radius r in the positive sense, then the real axis from r to ε, and, finally, the circle $\gamma(\varepsilon)$ of centre 0 and radius ε in the negative sense (cf. figure 8). The integral

$$\int_{\delta(r,\varepsilon)} \frac{R(z)}{z^\alpha}\,dz$$

is equal to the sum of the residues of the poles of $\frac{R(z)}{z^\alpha}$ contained in D, if r has been chosen sufficiently large and ε sufficiently small. We have

$$\int_{\delta(r,\varepsilon)} \frac{R(z)}{z^\alpha}\,dz = \int_{\gamma(r)} \frac{R(z)}{z^\alpha}\,dz + \int_{\gamma(\varepsilon)} \frac{R(z)}{z^\alpha}\,dz + (1 - e^{-2\pi i\alpha}) \int_\varepsilon^r \frac{R(x)}{x^\alpha}\,dx$$

because, when the argument of z is equal to 2π, we have $z^\alpha = e^{2\pi i\alpha}|z|^\alpha$. Since the argument of z remains bounded, $zf(z)$ tends to 0 when z tends

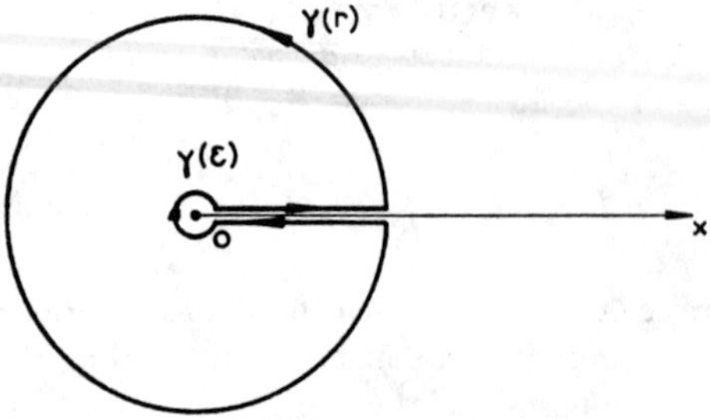

Fig. 8.

to 0 or when $|z|$ tends to infinity; thus the integrals along $\gamma(r)$ and $\gamma(\varepsilon)$ tend to 0 as r tends to ∞ and ε tends to 0 (lemmas 1 and 2). On the limit, we have

$$(4.\ 2) \qquad (1 - e^{-2i\pi\alpha})I = 2\pi i \sum \operatorname{Res}\left(\frac{R(z)}{z^\alpha}\right),$$

and this relation allows us to calculate I.

Example. To evaluate $I = \int_0^{+\infty} \frac{dx}{x^\alpha(1+x)}$, $(0 < \alpha < 1)$. Here we have $R(z) = \frac{1}{1+z}$; there is only one pole $z = -1$; because the branch of the argument of z is equal to π at this point, the residue of $\frac{R(z)}{z^\alpha}$ at this pole is equal to $\frac{1}{e^{\pi i\alpha}}$. Relation (4. 2) then gives

$$I = \frac{\pi}{\sin \pi\alpha}.$$

5 *th type.* Let us consider integrals of the form

$$\int_0^{+\infty} R(x) \log x \, dx,$$

where R is a rational function with no pole on the positive real axis $x \geqslant 0$, and such that $\lim_{x \to +\infty} xR(x) = 0$. This last condition ensures that the integral is convergent.

We consider the same open set D as for integrals of the 4th type and the same path of integration. Here again, we must specify the branch chosen for $\log z$; we shall choose the argument of z between 0 and 2π. For a reason which will soon be apparent, we shall integrate, not the function $R(z) \log z$, but the function $R(z)(\log z)^2$. Here again the integrals along the circles $\gamma(r)$ and $\gamma(\varepsilon)$ tend to 0 as r tends to ∞, and ε tends to 0 because of lemmas 1 and 2. When the argument of z is equal to 2π, we have

$$\log z = \log x + 2\pi i,$$

x denoting the modulus of z. Thus we have the relation

$$\int_0^{+\infty} R(x)(\log x)^2 dx - \int_0^{+\infty} R(x)(\log x + 2\pi i)^2 dx = 2\pi i \sum \text{Res}\{R(z)(\log z)^2\};$$

hence,

$$(5.1) \quad -2\int_0^{+\infty} R(x)\log x\, dx - 2\pi i \int_0^{+\infty} R(x)\, dx = \sum \text{Res}\{R(z)(\log z)^2\}.$$

Basically this only gives one relation between the two integrals $\int_0^{+\infty} R(x)\, dx$ and $\int_0^{+\infty} R(x)\log x\, dx$. Let us suppose, however, that the rational function R is *real* (that is, it takes real values for x real); by separating real and imaginary parts of the relation (5.1), we obtain the two relations

$$(5.2) \quad \int_0^{+\infty} R(x)\log x\, dx = -\frac{1}{2}\text{Re}\left(\sum \text{Res}\{R(z)(\log z)^2\}\right),$$

$$(5.3) \quad \int_0^{+\infty} R(x)\, dx = -\frac{1}{2\pi}\text{Im}\left(\sum \text{Res}\{R(z)(\log z)^2\}\right).$$

The summation extends over all the poles of the rational function $R(z)$ contained in D.

Example. To evaluate the integral

$$I = \int_0^{+\infty} \frac{\log x}{(1+x)^3}\, dx.$$

The residue of $\frac{(\log z)^2}{(1+z)^3}$ at the pole $z = -1$ is equal to the coefficient of t^2 in the limited expansion of $(i\pi + \log(1-t))^2$; it is therefore $1 - i\pi$, and we find $I = -\frac{1}{2}$.

Note. By integrating the function $R(z)\log z$ in the same way, we obtain the formula

$$(5.4) \quad \int_0^{+\infty} R(x)\, dx = -\sum \text{Res}\{R(z)\log z\}.$$

The above method can also be applied in some cases when the rational function R has a *simple pole* at $x = 1$; in this case, the integral $\int_0^{+\infty} R(x)\log x\,dx$ still has a meaning because the principal branch of $\log z$ has a simple zero at the point 1. It is then necessary to modify the contour of integration which we used before; when we integrate along the positive real axis with the argument of z equal to 2π, we must by-pass

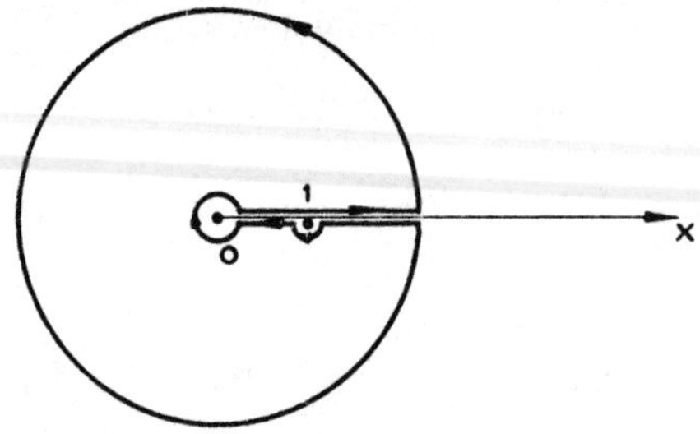

Fig. 9.

the point $z = 1$ along a semi-circle of centre 1 and small radius (fig. 9). The reader should prove that, when the function R is real, it satisfies

$$(5.5)\quad \int_0^{+\infty} R(x)\log x\,dx = \pi^2\,\mathrm{Re}\,(\mathrm{Res}\,(R, 1)) - \frac{1}{2}\mathrm{Re}\left(\sum \mathrm{Res}\,(f)\right),$$

where f denotes the function $R(z)\,(\log z)^2$ and where the summation extends over all the poles of f other than $z = 1$. For example, it can be verified that

$$\int_0^{+\infty} \frac{\log x}{x^2 - 1}\,dx = \frac{\pi^2}{4}.$$

Exercises

1. Let $f(z)$ be holomorphic in $|z| < R$, $R > 1$. Evaluate the integrals

$$\int_{|z|=1} \left(2 \pm \left(z + \frac{1}{z}\right)\right)\frac{f(z)}{z}\,dz$$

taken over the unit circle in the positive sense in two different ways and deduce the following relations :

$$\begin{cases} \dfrac{2}{\pi}\displaystyle\int_0^{2\pi} f(e^{i\theta})\cos^2\frac{\theta}{2}\,d\theta = 2f(0) + f'(0), \\ \dfrac{2}{\pi}\displaystyle\int_0^{2\pi} f(e^{i\theta})\sin^2\frac{\theta}{2}\,d\theta = 2f(0) - f'(0). \end{cases}$$

2. Let $f(z)$ be a holomorphic function in an open set containing the disc $|z| \leqslant R$ and let γ be the image of the circle $|z| = R$ under the mapping $z \to f(z)$; suppose that f is simple, i.e. $f(z) \neq f(z')$ if $z \neq z'$. Show that the length L of γ is equal to $R \int_0^{2\pi} |f'(Re^{i\theta})|\, d\theta$; deduce that

$$L \geqslant 2\pi R |f'(0)|.$$

Show that, under the same conditions, the area A of the image of the closed disc $|z| \leqslant R$ under the same mapping is given by

$$A = \iint_{|z| \leqslant R} |f'(x + iy)|^2\, dx\, dy;$$

deduce the inequality

$$A \geqslant \pi R^2 |f'(0)|^2.$$

(Change to polar coordinates and note that, for $0 \leqslant r \leqslant R$,

$$|f'(0)|^2 = \frac{1}{4\pi^2} \left| \int_0^{2\pi} f'(re^{i\theta})\, d\theta \right|^2$$
$$\leqslant \frac{1}{4\pi^2} \int_0^{2\pi} |f'(re^{i\theta})|^2\, d\theta \int_0^{2\pi} d\theta = \frac{1}{2\pi} \int_0^{2\pi} |f'(re^{i\theta})|^2\, d\theta,$$

because of the Cauchy-Schwarz inequality for integrals.)

3. Show that, if $f(z)$ is holomorphic in an open set containing the closed disc $|z| \leqslant 1$, then

$$\frac{1}{2\pi i} \int_{|z|=1} \frac{\overline{f(z)}}{z - a}\, dz = \begin{cases} \overline{f(0)} & \text{if} \quad |a| < 1, \\ \overline{f(0)} - \overline{f(1/\bar{a})} & \text{if} \quad |a| > 1, \end{cases}$$

where the integral is taken in the positive sense. (Use exercise 1.b) of chapter II and Cauchy's integral formula.)

4. Let $f(z)$ be a holomorphic function in the whole plane, and suppose that there is an integer n and two positive real numbers R and M such that

$$|f(z)| \leqslant M . |z|^n \qquad \text{for} \qquad |z| \geqslant R.$$

Show then that $f(z)$ is a polynomial of degree at most n.

5. Let f be a non-constant, holomorphic function in a connected open set D, and let D′ be a connected open set whose closure $\overline{D}'$ is a compact subset of D. Show that, if $|f(z)|$ is constant on the frontier of D′, there is at least one zero of $f(z)$ in D′. (Use *reductio ad absurdum* by considering $1/f(z)$.)

6. Let D be a bounded, connected, open set and consider n points P_1, $P_2, \ldots, P_n$ in the plane $\mathbf{R}^2$. Show that the product $\overline{PP_1}. \overline{PP_2} \ldots \overline{PP_n}$ of the distances from a point P, which varies in the closure $\overline{D}$, to the points $P_1, P_2, \ldots, P_n$, attains its maximum at a frontier point of D.

7. Let $f(z)$ be a holomorphic function in the disc $|z| < R_1$ and put $M(r) = \sup_{|z|=r} |f(z)|$, for $0 \leqslant r < R$. Show that

a) $M(r)$ is a continuous, monotonic increasing (in the broad sense), function of r in $0 \leqslant r < R$,

b) if $f(z)$ is not constant, $M(r)$ is strictly increasing.

8. Hadamard's three circles theorem : let $f(z)$ be a holomorphic function in an open set containing the closed annulus

$$r_1 \leqslant |z| \leqslant r_2 \qquad (0 < r_1 < r_2),$$

and put $M(r) = \sup_{|z|=r} |f(z)|$ for $r_1 < r < r_z$. Show that the following inequality holds

$$M(r) \leqslant M(r_1)^{\frac{\log r_2 - \log r}{\log r_2 - \log r_1}} . M(r_2)^{\frac{\log r - \log r_1}{\log r_2 - \log r_1}}, \tag{1}$$

for $r_1 \leqslant r \leqslant r_2$. (Apply the maximum modulus principle to the function $z^p(f(z))^q$ with p, q integers and $q > 0$; choose α real such that $r_1^\alpha M(r_1) = r_2^\alpha M(r_2)$, and a sequence of pairs of integers (p_n, q_n) such that $\lim_{n\to\infty} p_n/q_n = \alpha$.) Verify that inequality (1) expresses that $\log M(r)$ is a convex function of $\log r$ for $r_1 \leqslant r \leqslant r_2$.

9. Let $f(z)$ be holomorphic in $|z| < R$ and put

$$I_2(r) = \frac{1}{2\pi}\int_0^{2\pi} |f(re^{i\theta})|^2 \, d\theta, \quad \text{for} \quad 0 \leqslant r < R.$$

Show that, if a_n denotes the n-th Talylor coefficient of $f(z)$ at the point $z = 0$, then

$$I_2(r) = \sum_{n\geqslant 0} |a_n|^2 r^{2n};$$

deduce that, if $0 \leqslant r < R$,

(i) $I_2(r)$ is a continuous, monotonic increasing (in the broad sense), function of r;

(ii) $|f(0)|^2 \leqslant I_2(r) \leqslant (M(r))^2$, ($M(r)$ has the same meaning as in 7.);

(iii) $\log I_2(r)$ is a convex function of $\log r$ in the case when f is not identically zero (show that, if

$$s = \log r, \quad J(s) = I_2(e^s) = \sum_{n\geqslant 0} |a_n|^2 e^{2ns}, \quad \text{then} \quad (\log J)'' = \frac{J''J - (J')^2}{J^2};$$

to show that $JJ'' - (J')^2 \geqslant 0$, use the Cauchy-Schwarz inequality for absolutely convergent series :

$$\left|\sum_{n\geqslant 0} \alpha_n\bar{\beta}_n\right|^2 \leqslant \left(\sum_{n\geqslant 0} |\alpha_n|^2\right)\left(\sum_{n\geqslant 0} |\beta_n|^2\right)\bigg).$$

10. Let f be a holomorphic function in the disc $|z| < 1$, such that $|f(z)| < 1$ in this disc; if there exist two distinct points a and b in the disc such that $f(a) = a$ and $f(b) = b$, show that $f(z) = z$ in the disc. (Consider the function $g(z) = \dfrac{h(z) - a}{1 - \bar{a}h(z)}$, with $h(z) = f\left(\dfrac{z+a}{1+\bar{a}z}\right)$, for which $g(0) = 0$, $g\left(\dfrac{b-a}{1-\bar{a}b}\right) = \dfrac{b-a}{1-\bar{a}b}$, and $|g(z)| < 1$ in the disc.)

11. Let f be a holomorphic function in an open set containing the disc $|z| \leqslant R$. For $0 \leqslant r \leqslant R$, put

$$A(r) = \sup_{0\leqslant\theta\leqslant 2\pi} \operatorname{Re}(f(re^{i\theta})).$$

(i) Show that $A(r)$ is a continuous, monotonic increasing (in the broad sense), function of r (note that $e^{\operatorname{Re} f(z)} = |e^{f(z)}|$).

(ii) Show that, if $f(0) = 0$ also, then, for $0 \leqslant r < R$,

$$M(r) \leqslant \frac{2r}{R-r} A(R).$$

(Consider the function $g(z) = f(z)/(2A(R) - f(z))$.)

(iii) Show that, for $0 \leqslant r < R_1$

$$M(r) \leqslant \frac{2r}{R-r} A(R) + \frac{R+r}{R-r}|f(0)|.$$

12. Let x be a complex parameter.

(i) Show that the Laurent expansion of the function

$$\exp\left(x\left(z + \frac{1}{z}\right)\Big/2\right)$$

at the origin $z = 0$, is of the following form :

$$\exp\left(x\left(z + \frac{1}{z}\right)\Big/2\right) = a_0 + \sum_{n\geqslant 1} a_n\left(z^n + \frac{1}{z^n}\right) \qquad \text{for} \qquad 0 < |z| < +\infty,$$

with

$$a_n = \frac{1}{\pi}\int_0^{\pi} e^{x\cos t}\cos nt\, dt, \qquad \text{for} \quad n \geqslant 0.$$

Show similarly that the function $\exp\left(x\left(z - \frac{1}{z}\right)/2\right)$ has the expansion

$$\exp\left(x\left(z - \frac{1}{z}\right)/2\right) = b_0 + \sum_{n\geq 1} b_n\left(z^n + \frac{(-1)^n}{z^n}\right) \quad \text{for} \quad 0 < |z| < +\infty,$$

with

$$b_n = \frac{1}{\pi}\int_0^{\pi} \cos(nt - x\sin t)\,dt, \quad \text{for} \quad n \geq 0.$$

(Note that, if $z' = -1/z$, then

$$\exp(x(z' - 1/z')/2) = \exp(x(z - 1/z)/2) \quad \text{for} \quad 0 < |z| < +\infty.)$$

(ii) Let m, n be two integers ≥ 0. Show that

$$\frac{1}{2\pi i}\int_{|z|=1} \frac{(z^2 \pm 1)^m\,dz}{z^{m+n+1}} = \begin{cases} \dfrac{(\pm 1)^p (n+2p)!}{p!\,(n+p)!}, & \text{if } m = n + 2p, \text{ with } p \text{ an integer} \geq 0, \\ 0 & \text{otherwise,} \end{cases}$$

and deduce the power series expansions of a_n, b_n as functions of the parameter x (b_n, as a function of x, is called Bessel's function of the first kind).

13. Let $f(z)$ be a meromorphic function in a neighbourhood of the origin $z = 0$ with a simple pole at the origin. Let x be any complex number. Show that the Laurent expansion of the function of z

$$\frac{f'(z)}{f(z) - x}$$

is of the form

$$-\frac{1}{z} + u_1 + u_2 z + \cdots + u_{n+1} z^n + \cdots,$$

where u_n is a polynomial in x of degree n. (An identification can be made by using the Taylor expansion of the function $zf(z)$.)

14. Let $f(z)$ be a holomorphic function in the upper half-plane P^+ defined by $\operatorname{Im}(z) > 0$, and suppose that $f(z+1) = f(z)$ for all $z \in P^+$. Show that there is a holomorphic function $g(t)$ in the punctured disc $0 < |t| < 1$, such that

$$f(z) = g(e^{2\pi i z}), \quad \text{for} \quad z \in P^+.$$

Deduce that $f(z)$ has an expansion of the form

$$f(z) = \sum_{-\infty < n < +\infty} a_n e^{2\pi i n z},$$

where

$$a_n = \int_0^1 f(x + iy)e^{-2\pi in(x+iy)}\,dx,$$

for any $y > 0$. Show that this series is normally convergent in any compact subset of P^+. Show also that, if there exists a constant $M > 0$ and an integer n_0 such that

$$|f(x + iy)| \leqslant Me^{2\pi n_0 y} \qquad \text{for all sufficiently large } y,$$

and uniformly in x, then the expansion is of the form

$$f(z) = \sum_{n \geqslant -n_0} a_n e^{2\pi inz}.$$

15. (i) Show that the function $f(z) = 1/(e^z - 1)$ is meromorphic in the whole plane $\mathbf{C}$ and has simple poles at the points $z = 2p\pi i$, p an integer. Calculate its Laurent expansion at the point $z = 2p\pi i$. If a_n $(n \geqslant -1)$ denote the coefficients of the expansion for $p = 0$, show that $a_{2q} = 0$ for $q = 1, 2, \ldots$, and if

$$B_n = (-1)^{n-1}(2n)!\, a_{2n-1}, \quad \text{for } n \geqslant 1,$$

show that the following recurrence relation holds :

$$\frac{1}{(2n+1)!} - \frac{1}{2(2n)!} + \sum_{1 \leqslant \nu \leqslant n} \frac{(-1)^{\nu-1} B_\nu}{(2\nu)!\,(2n - 2\nu + 1)!} = 0,$$

for $n \geqslant 1$ $\Bigl($by equating coefficients on the two sides of the relation

$$\Bigl(a_{-1}/z + \sum_{n \geqslant 0} a_n z^n\Bigr)\Bigl(\sum_{m \geqslant 1} z_m/m!\Bigr) = 1\Bigr).$$

(ii) For $n \geqslant 1$, put

$$f_{2n}(z) = \frac{1}{z^{2n}(e^z - 1)},$$

and let γ_m be the perimeter of the square whose vertices have complex coordinates $\pm (2m + 1)\pi \pm (2m + 1)\pi i$. Show that

$$|f_{2n}(z)| \leqslant 2/((2m + 1)\pi)^{2n} \qquad \text{if } z \text{ is on } \gamma_m,$$

and deduce, by integrating $f_{2n}(z)$ round the contour γ_m in the positive sense and letting $m \to \infty$, that

$$\sum_{p \geqslant 1} 1/p^{2n} = \frac{(2\pi)^{2n} B_n}{2(2n)!}.$$

(N.B. The numbers B_n are called the Bernoulli numbers.)

16. Let c be an essential singularity of a holomorphic function $f(z)$ in the punctured disc D given by $0 < |z - c| < \rho$.

(i) For any $\gamma \in \mathbf{C}$ and $\varepsilon > 0$, show that there exists a $z' \in \mathrm{D}$ and a real number $\varepsilon' > 0$ such that

$$\bar{\Delta}(f(z'), \varepsilon') \subset \Delta \cap \Delta(\gamma, \varepsilon),$$

where Δ denotes the image of D under the mapping $z \to f(z)$ and where $\Delta(b, r)$ (resp. $\bar{\Delta}(b, r)$) is the open (resp. closed) disc of radius r centred at b (note that proposition 4. 2 of § 5 implies that Δ is *open* (this also follows from the theorem in chapter VI, § 1, no. 3), and use Weierstrass' theorem, no. 4 of § 4).

(ii) Let D be the punctured disc $0 < |z - c| < \rho/2^n$ and let Δ_n be its image under f, for $n \geqslant 0$. Given $\gamma_0 \in \mathbf{C}$, $\varepsilon_0 > 0$, show, by induction on n, the existence of a sequence $(\varepsilon_n)_{n \geqslant 1}$ of positive real numbers and a sequence $(z_n)_{n \geqslant 1}$ of points of D satisfying the following conditions :

$$z_n \in \mathrm{D}_{n-1}, \qquad \varepsilon_0 > \varepsilon_1 > \varepsilon_2 > \ldots, \qquad \bar{\Delta}(f(z_1), \varepsilon_1) \subset \Delta \cap \Delta(\gamma_0, \varepsilon_0)$$
$$\bar{\Delta}(f(z_{n+1}), \varepsilon_{n+1}) \subset \Delta_n \cap \Delta(f(z_n), \varepsilon_n) \qquad \text{for} \qquad n \geqslant 1,$$

deduce that there exists a sequence $(c_n)_{n \geqslant 0}$ of points in D such that

$$\lim c_n = c \qquad \text{and} \qquad f(c_n) = \gamma \qquad \text{for all } n, \text{ with} \qquad |\gamma - \gamma_0| < \varepsilon_0,$$

and that $f(z)$ is not simple in any punctured disc $0 < |z - c| < r$ however small r is.

17. Let $\varphi : (x, y, u) \to z$ be the stereographic projection of $\mathbf{S}_2 - \mathrm{P}$ onto $\mathbf{C}$.

(i) Express x, y and u as functions of z.

(ii) Show that, if C is a circle of $\mathbf{S}_2$, which does not pass through P, $\varphi(\mathrm{C})$ is a circle in the plane $\mathbf{C}$, and that, if C passes through P, $\varphi(\mathrm{C} - \mathrm{P})$ is a line in $\mathbf{C}$.

(iii) Let $z_1, z_2 \in \mathbf{C}$; show that a necessary and sufficient condition for $\varphi^{-1}(z_1)$ and $\varphi^{-1}(z_2)$ to be antipodal is that $z_1 \bar{z}_2 = -1$.

(iv) Show that the distance $\overline{\mathrm{P}_1\mathrm{P}_2}$ (in $\mathbf{R}^3$) between

$$\mathrm{P}_1 = \varphi^{-1}(z_1) \qquad \text{and} \qquad \mathrm{P}_2 = \varphi^{-1}(z_2)$$

is given by the formula

$$\overline{\mathrm{P}_1\mathrm{P}_2} = \frac{2|z_1 - z_2|}{\sqrt{(1 + |z_1|^2)(1 + |z_2|^2)}}.$$

What happens to the formula when z_2 tends towards the point infinity?

18. Show that a meromorphic function on the Riemann sphere is necessarily rational. (Show first that such a function can only have a finite number of poles.)

19. Rouché's theorem. Let $f(z)$ and $g(z)$ be holomorphic functions in an open set D and let $\Gamma = (\Gamma_i)_{i \in I}$ be the oriented boundary of a compact subset K of D. If

$$|f(z)| > |g(z)| \quad \text{on} \quad \Gamma,$$

show that the number of zeros of $f(z) + g(z)$ in K is equal to the number of zeros of $f(z)$ in K. (Consider the closed paths $f \circ \Gamma_i$, $i \in I$ and apply proposition 4.1 of § 5 and proposition 8.3 of chapter II, §1.)

Example. If $f(z)$ is holomorphic in an open set containing the closed disc $|z| \leqslant 1$ and if $|f(z)| < 1$ for $|z| = 1$, then the equation $f(z) = z^n$ has exactly n solutions in $|z| < 1$, for any integer $n \geqslant 0$.

20. Evaluate the following integrals by calculating residues:

(i) $\displaystyle\int_0^{+\infty} \frac{dx}{(a+bx^2)^n}$ $(a, b > 0)$, (ii) $\displaystyle\int_0^{+\infty} \frac{\cos 2ax - \cos 2bx}{x^2}\, dx$ (a, b real),

(iii) $\displaystyle\int_0^{+\infty} \frac{x^2 - a^2}{x^2 + a^2} \frac{\sin x}{x}\, dx$ $(a > 0)$, (iv) $\displaystyle\int_0^{\pi} \frac{\cos nt\, dt}{1 - 2a\cos t + a^2}$ $(|a| \neq 1)$

(integrate the function $z^n/(z-a)(z-1/a)$ round the unit circle).

21. Integrate the function $f(z) = \dfrac{1}{(z^2+a^2)\log z}$, where log denotes the branch such that $-\pi \leqslant \arg z \leqslant \pi$, along the closed path $\delta(r, \varepsilon)$ defined as follows: describe in turn the negative real axis from $-r$ to $-\varepsilon$, then the circle $\gamma(\varepsilon)$ of radius ε centred at 0 in the negative sense, then the negative real axis from $-\varepsilon$ to $-r$, and, finally, the circle $\gamma(r)$ of radius r centred at 0 anticlockwise $(0 < \varepsilon < a < r)$; deduce that

$$\int_0^{\infty} \frac{dx}{(x^2+a^2)((\log x)^2 + \pi^2)} = \frac{\pi}{2a((\log a)^2 + \pi^2/4)} - \frac{1}{1+a^2}.$$

22. Let a be > 0 and ν be real. Show that

$$\int_0^{\infty} \frac{\cos \nu x\, dx}{\cosh x + \cosh a} = \frac{\pi \sin \nu a}{\sinh \pi\nu \sinh a},$$

by integrating the function $e^{i\nu z}/(\cosh z + \cosh a)$ along the perimeter of the rectangle with vertices $\pm R$, $\pm R + 2\pi i$.

23. (i) Let n be an integer $\geqslant 2$. Show that

$$\int_0^{\infty} \frac{dx}{1+x^n} = \frac{\pi/n}{\sin(\pi/n)},$$

by integrating the function $1/(1+z^n)$ along the contour formed by the segment $[0, R]$ of the positive real axis, the arc represented by Re^{it}, $0 \leqslant t \leqslant 2\pi/n$, and the segment represented by $re^{2\pi i/n}$, $0 \leqslant r \leqslant R$.

(ii) Let n be an integer $\geqslant 2$ and let α be a real number such that $n > 1 + \alpha > 0$. Evaluate, by the same method, the integral

$$\int_0^\infty \frac{x^\alpha\, dx}{1 + x^n}.$$

24. Let p, q be two real numbers > 0 and let n be an integer $\geqslant 1$. By integrating the function $z^{n-1}e^{-z}$ along a contour analogous to the above (in exercise 23), but with a suitable choice of the angle at the origin, prove the following relations :

$$\int_0^\infty x^{n-1}e^{-px}\cos qx\, dx = \frac{(n-1)!\,\mathrm{Re}\,(p+iq)^n}{(p^2+q^2)^n},$$

$$\int_0^\infty x^{n-1}e^{-px}\sin qx\, dx = \frac{(n-1)!\,\mathrm{Im}\,(p+iq)^n}{(p^2+q^2)^n}.$$

$\left(\text{Recall that } \int_0^\infty x^{n-1}e^{-x}\, dx = (n-1)!.\right)$

25. (i) Show that the function $\pi \cot \pi z$ is meromorphic in the whole complex plane, that it has simple poles at the points $z = n$ for n an integer, and that its residue at the pole $z = n$ is equal to 1 for all n. Let

$$f(z) = P(z)/Q(z)$$

be a rational function such that deg $Q >$ deg. $P + 1$, and let $a_1, a_2, \ldots, a_m$ be its poles and let $b_1, b_2, \ldots, b_m$ be the corresponding residues. Suppose also that the a_q are not integers for $1 \leqslant q \leqslant m$. Let γ_n denote the perimeter of the square with vertices $\pm\left(n+\frac{1}{2}\right) \pm \left(n+\frac{1}{2}\right)i$, where n is a positive integer. Show that there exist two positive real numbers M_1, K independent of n such that

a) $\quad |\pi \cot \pi z| \leqslant M_1 \quad$ on $\quad \gamma_n$

b) $\quad |f(z)| \leqslant K/|z|^2 \quad$ for sufficiently large $|z|$.

Deduce that

$$\lim_{n\to\infty} \int_{\gamma_n} f(z)\pi \cot \pi z\, dz = 0.$$

and that

$$(1) \qquad \lim_{n\to\infty} \sum_{-n\leqslant p\leqslant n} f(p) = -\sum_{1\leqslant q\leqslant m} b_q \pi \cot \pi a_q.$$

(Note : *b*) implies that $\lim_{n, n' \to \infty} \sum_{-n \leqslant p \leqslant n'} f(p)$ exists, thus the left hand sidef of (1) can be replaced by $\sum_{-\infty < p \leqslant \infty} f(p)$.)

Example. $\sum_{n \geqslant 1} 1/(a + bn^2)$, $\sum_{n \geqslant 1} n^2/(n^4 + a^4)$ (a, b positive real numbers).

(ii) Show that the conclusion is valid even if we only have deg Q > deg P. (Show first that $f(z)$ can be written $g(z) + c/z$ with c a constant and $g(z)$ a rational function which satisfies the conditions of (i); show next that $\int_{\gamma_n} \frac{\cot \pi z}{z} dz = 0$ (the integrals along opposite sides cancel). Note : $\lim_{n, n' \to \infty} \sum_{-n \leqslant p \leqslant n'} f(p)$ does not exist in this case.)

Example. Calculate $\lim_{n \to \infty} \sum_{-n \leqslant p \leqslant n} \frac{1}{x - p}$, and deduce the value o $\sum_{p \geqslant 1} \frac{1}{x^2 - p^2}$ when x is not an integer.

(iii) Let α be a real number such that $-\pi < \alpha < \pi$. Show that : *c*) there exists a positive real number M_2, which does not depend on n, such that

$$\left|\frac{e^{i\alpha z}}{\sin \pi z}\right| \leqslant M_2 \quad \text{on} \quad \gamma_n,$$

d)

$$\lim_{n \to \infty} \int_{\gamma_n} \frac{e^{i\alpha z}}{z \sin \pi z} dz = 0.$$

(Note that

$$\int_{\gamma_n} \frac{e^{i\alpha z}}{z \sin \pi z} dz = 2i \int_{\gamma'_n} \frac{\sin \alpha z}{z \sin \pi z} dz + 2i \int_{\gamma''_n} \frac{\sin \alpha z}{z \sin \pi z} dz,$$

where γ'_n (resp. γ''_n) denotes the line segment represented by

$$z = n + \frac{1}{2} + iy, \quad |y| \leqslant n + \frac{1}{2} \left(\text{resp. } z = x + i\left(n + \frac{1}{2}\right), \ |x| \leqslant n + \frac{1}{2}\right),$$

and use exercise 14. of chapter 1.) Deduce finally that, if $f(z)$ is a rational function and satisfies the conditions of (ii), then

$$\lim_{n \to \infty} \sum_{-n \leqslant p \leqslant n} (-1)^p f(p) e^{i\alpha p} = -\pi \sum_{1 \leqslant q \leqslant m} b_q \frac{e^{i\alpha a_q}}{\sin \pi a_q}.$$

Example. Take $f(z) = 1/(x - z)$ and show that, if $-\pi < \alpha < \pi$, then

$$\begin{cases} \sum_{n \geqslant 1} (-1)^n \frac{\cos \alpha n}{x^2 - n^2} = \frac{\pi}{2x} \frac{\cos \alpha x}{\sin \pi x} - \frac{1}{2x^2}, \\ \sum_{n \geqslant 1} (-1)^n \frac{n \sin \alpha n}{x^2 - n^2} = \frac{\pi}{2} \frac{\sin \alpha x}{\sin \pi x}, \end{cases}$$

for $x \neq 0, \pm 1, \pm 2, \ldots$.

CHAPTER IV

Analytic Functions of Several Variables; Harmonic Functions

1. Power Series in Several Variables

In what follows, we shall only discuss the case of two variables so as not to let the notation become too complicated; however, the arguments go over to the case of any finite number of variables without difficulty.

1. The algebra K[[X, Y]]

A formal power series in X and Y with coefficients in a field K is an expression of the form $S(X, Y) = \sum_{p, q \geqslant 0} a_{p,q} X^p Y^q$, where the coefficients $a_{p,q}$ belong to the field K.

We define, as in chapter 1, § 1, addition of two formal power series and multiplication of a formal power series by a scalar. The set K[[X, Y]] of formal series thus has a vector space structure over the field K. We define the product of two formal power series, and K[[X,Y]] becomes an algebra.

The *order* of a formal power series which is not identically zero is defined to be the smallest integer n such that

$$\sum_{p+q=n} a_{p,q} X^p Y^q \neq 0.$$

It can be shown that the order of the product of two non-zero series is equal to the sum of the orders of these series; in particular, K[[X, Y]] is an integral domain.

We shall not develop the theory of substituting formal power series for the variables X and Y; the theory does not present any special difficulties, but the series that are substituted must have order $\geqslant 1$. As an exercise, the reader is invited to prove a proposition similar to proposition 5.1 of § 1 of chapter 1.

2. DOMAIN OF CONVERGENCE OF A MULTIPLE POWER SERIES

We suppose from now on that the field K is either **R** or **C**. As in chapter I, § 2, no. 3, we associate with each formal series

$$\sum_{p,\, q \geqslant 0} a_{p,\, q} X^p Y^q$$

the series of positive (or zero) terms

$$\sum_{p,\, q \geqslant 0} |a_{p,\, q}| (r_1)^p (r_2)^q,$$

where r_1 and r_2 are real variables $\geqslant 0$.

Let Γ be the set of points (r_1, r_2) of the quadrant $r_1 \geqslant 0$, $r_2 \geqslant 0$ of the plane such that $\sum_{p,\, q} |a_{p,\, q}| (r_1)^p (r_2)^q < +\infty$. The series $\sum_{p,\, q} a_{p,\, q} (z_1)^p (z_2)^q$ is then *absolutely convergent* for any pair of (real or complex) numbers z_1 and z_2 such that $|z_1| \leqslant r_1$, $|z_2| \leqslant r_2$. The set Γ is not empty because it obviously contains the origin (0, 0).

Definition. The *domain of convergence* of the series S(X, Y) is defined to be the set Δ of points of the quadrant $r_1 \geqslant 0$, $r_2 \geqslant 0$ *interior to* Γ.

The domain of convergence is then an open set of the quadrant. This set can be empty : it is, in fact, quite easy to construct an example where Γ consists only of the origin.

If we apply this definition to the case of a single variable z, we see that the domain of convergence is merely the interval $]0, \rho[$ where ρ is the radius of convergence of the power series.

PROPOSITION 2. 1. *A necessary and sufficient condition that* $(r_1, r_2) \in \Delta$ *is that there exist* $r_1' > r_1$, $r_2' > r_2$ *such that* $(r_1', r_2') \in \Gamma$.

The condition is necessary, since the series $\sum_{p,q} |a_{p,q}| (r_1')^p (r_2')^q$ must converge at all points sufficiently near to (r_1, r_2). It is sufficient because Γ then contains all the points (ρ_1, ρ_2) such that $\rho_1 \leqslant r_1'$, $\rho_2 \leqslant r_2'$ and consequently the point (r_1, r_2) is interior to Γ.

In particular, a necessary and sufficient condition for the domain of convergence Δ to be non-empty is that there exists at least one pair (r_1, r_2) of numbers > 0 such that

$$\sum_{p,q} |a_{p,q}| (r_1)^p (r_2)^q < +\infty.$$

PROPOSITION 2. 2. *If* (r_1, r_2) *belongs to the domain of convergence, the series* $S(z_1, z_2)$ *converges normally for* $|z_1| \leqslant r_1$, $|z_2| \leqslant r_2$. *If* $(|z_1|, |z_2|)$ *does not belong to the closure of* Γ, *the series* $S(z_1, z_2)$ *is divergent.*

The proof depends, as in the case of series in one variable, on Abel's lemma :

LEMMA. *If* $|a_{p,q}|(r_1')^p(r_2')^q \leqslant M$ (M *independent of* p *and* q) *and if* $r_1 < r_1'$ *and* $r_2 < r_2'$, *then the series* $\sum_{p,q} a_{p,q}(z_1)^p(z_2)^q$ *converges normally for* $|z_1| \leqslant r_1$, $|z_2| \leqslant r_2$.

This lemma is easily proved by bounding above the absolute values of the terms of the series by the terms of a double geometric progression. The reader is left to deduce for himself proposition 2. 2 from Abel's lemma.

By *abus de langage*, we also use 'domain of convergence' to denote the set of pairs (z_1, z_2) such that $(|z_1|, |z_2|)$ belongs to the domain of convergence Δ. Thus, for a single complex variable z, the domain of convergence is the open disc $|z| < \rho$, where ρ is the radius of convergence.

3. Operations on convergent power series

PROPOSITION 3. 1 (Addition and multiplication of power series). *Let* D *be an open set contained in the domain of convergence of the series* A(X, Y) *and in that of series* B(X, Y). *Then* D *is contained in the domain of convergence of each the series*

$$S(X, Y) = A(X, Y) + B(X, Y), \qquad P(X, Y) = A(X, Y)B(X, Y).$$

Moreover, if $(|z_1|, |z_2|) \in D$, *then*

$$S(z_1, z_2) = A(z_1, z_2) + B(z_1, z_2), \qquad P(z_1, z_2) = A(z_1, z_2)B(z_1, z_2).$$

The proof is analogous to that given in the case of series in one variable.

We define the partial derivatives of a power series

$$S(X, Y) = \sum_{p,q} a_{p,q}X^pY^q$$

in the obvious way :

$$\begin{cases} \dfrac{\partial S}{\partial X} = \sum_{p,q} pa_{p,q}X^{p-1}Y^q, \\ \dfrac{\partial S}{\partial Y} = \sum_{p,q} qa_{p,q}X^pY^{q-1}. \end{cases}$$

PROPOSITION 3.2. *The series* $\dfrac{\partial S}{\partial X}$ *has the same domain of convergence as the series* S. *When* $(|z_1|, |z_2|)$ *is in this domain, the function* $\dfrac{\partial S}{\partial X}(z_1, z_2)$ *is the partial derivative (with respect to the real or complex variable* z_1*) of the function* $S(z_1, z_2)$.

The proof is modelled on that of proposition 7.1 of chapter 1, § 2.

By successive differentiation, it can be shown that

$$(3.1) \qquad a_{p,q} = \frac{1}{p!\,q!}\,\frac{\partial^{p+q}S(0, 0)}{\partial z_1^p\,\partial z_2^q}.$$

2. Analytic Functions

Here we consider functions of several real or complex variables defined in an open set D. For simplification, we shall confine our attention to functions of two variables.

1. Functions with power series expansions

Definition. We say that a function $f(x, y)$ defined in a neighourhood of a point (x_0, y_0) *has a power series expansion* at the point (x_0, y_0) if there exists a formal power series S(X, Y) whose domain of convergence is not empty and such that

$$f(x, y) = \mathrm{S}(x - x_0, y - y_0)$$

for sufficiently small $|x - x_0|$ and $|y - y_0|$.

The power series S, if it exists, is *unique* because of formula (3. 1) of § 1. The same reasoning as in chapter 1, § 4 gives the following properties: If $f(x,y)$ has a power series expansion at the point (x_0, y_0), then the function f is infinitely differentiable in some neighbourhood of (x_0, y_0). The product fg of two functions f and g, which both have a power series expansion at the point (x_0, y_0), has a power series expansion at the point (x_0, y_0); if this product is identically zero in a neighbourhood of (x_0, y_0), then at least one of the functions f and g is identically zero in some neighourhood of (x_0, y_0).

2. Analytic functions; operations on these functions

Definition. A real- or complex-valued function $f(x, y)$ defined in an open set D is said to be *analytic* in D, if, for any point $(x_0, y_0) \in \mathrm{D}$, the function $f(x, y)$ has a power series expansion at the point (x_0, y_0).

We shall confine ourselves to stating the following properties without proofs : The analytic functions in an open set D form a ring and even an algebra. If $f(x,y)$ is analytic in D, $1/f(x,y)$ is analytic at any point $(x_0, y_0) \in \mathrm{D}$ where $f(x_0, y_0) \neq 0$. Any analytic function in D is infinitely differentiable, and its derivatives are analytic functions in D. The composite of two analytic functions is analytic : precisely, if $f(x, y, z)$ is analytic in D and if $g_1(u, v)$, $g_2(u, v)$, $g_3(u, v)$ are analytic in an open set D′ and take their values in D, then the composed function $f(g_1(u, v), g_2(u, v), g_3(u, v))$ is analytic in D′.

PROPOSITION 2. 1. *The sum of a multiple power series is an analytic function of its variables in its domain of convergence.*

The proof is similar to that of proposition 2. 1 of § 4 of chapter 1. The reader should formulate a proposition similar to proposition 2.2 of the same paragraph.

3. THE PRINCIPLE OF ANALYTIC CONTINUATION

THEOREM. *Let $f(x, y)$ be an analytic function in a connected open set D and let $(x_0, y_0) \in$ D. The following conditions are equivalent:*

a) *f and all its derivatives vanish at (x_0, y_0);*

b) *f is identically zero in some neighbourhood of (x_0, y_0);*

c) *f is identically zero in D.*

The proof is modelled on that of the theorem of chapter 1, § 4, no. 3.

COROLLARY 1. *The ring of analytic functions in a connected open set D is an integral domain.*

COROLLARY 2. (principle of analytic continuation). *If two analytic functions f and g in a connected open set D coincide in some neighbourhood of a point of D, then they are identical in D.*

3. Harmonic Functions of Two Real Variables

1. DEFINITION OF HARMONIC FUNCTIONS

Definition. A function of $f(x, y)$ of two real variables x and y defined in an open set D is said to be *harmonic* in D if it has continuous derivatives of the second order and satisfies the condition

$$\frac{\partial^2 f}{\partial x^2} + \frac{\partial^2 f}{\partial y^2} = 0. \tag{1. 1}$$

The differential operator $\frac{\partial^2}{\partial x^2} + \frac{\partial^2}{\partial y^2}$ is called the Laplacian and is often denoted by Δ.

A harmonic function f of n real variables $x_1, \ldots, x_n$ is defined to be a function with continuous derivatives of the second order which satisfy

$$\frac{\partial^2 f}{\partial x_1^2} + \cdots + \frac{\partial^2 f}{\partial x_n^2} = 0;$$

but the following only applies to the case of two variables.

Let us introduce the differentiations $\frac{\partial}{\partial z}$ and $\frac{\partial}{\partial \bar{z}}$ with respect to the complex variables $z = x + iy$ and its conjugate $\bar{z} = x - iy$ (cf. chapter II, § 2, n° 3). Then

(1. 2) $$\frac{\partial^2}{\partial x^2} + \frac{\partial^2}{\partial y^2} = 4\frac{\partial^2}{\partial z\, \partial \bar{z}},$$

and, consequently, condition (1. 1) is equivalent to the following :

(1. 3) $$\frac{\partial^2 f}{\partial z\, \partial \bar{z}} = 0.$$

Condition (1. 3) expresses then that f is a harmonic function.

Note. We consider both real- and complex-valued harmonic functions. By (1. 1), a necessary and sufficient condition for a complex-valued function $f = P + iQ$ (P and Q being real-valued) to be harmonic is that P and Q are harmonic. We shall often denote P by $\text{Re}(f)$ and Q by $\text{Im}(f)$.

2. Harmonic functions and holomorphic functions

Proposition 2. 1 *Any holomorphic function is harmonic.*

For, if f is holomorphic, it is infinitely differentiable, and, by taking the derivative $\frac{\partial}{\partial z}$ of the relation $\frac{\partial f}{\partial \bar{z}} = 0$, we obtain relation (1. 3).

Corollary. *The real and imaginary parts of a holomorphic function are harmonic functions.*

For example, $\log |z|$ is a harmonic function in the whole plane excluding the origin; for, in some neighbourhood of each point $z \neq 0$, $\log z$ has a branch, and $\log |z|$ is the real part of such a branch.

Proposition 2. 2. *Any real harmonic function $g(x, y)$ in an open set D is, in a neighbourhood of each point of D, the real part of a holomorphic function f which is determined up to addition of a constant.*

Proof. Since g is harmonic $\frac{\partial^2 g}{\partial z\, \partial \bar{z}} = 0$, and consequently $\frac{\partial g}{\partial z}$ is holomorphic in D. The differential form $2\frac{\partial g}{\partial z}dz$ has therefore a primitive f *locally*; in other words, in a neighbourhood of each point of D, there exists a function f, determined up to addition of a constant, such that

(2. 1) $$df = 2\frac{\partial g}{\partial z}dz.$$

This relation shows that f is holomorphic. Taking the complex conjugate of relation (2. 1) gives

$$(2.2) \qquad d\bar{f} = 2\frac{\partial g}{\partial \bar{z}}\, d\bar{z},$$

because, g being *real function*, the functions $\frac{\partial g}{\partial z}$ and $\frac{\partial g}{\partial \bar{z}}$ are complex conjugates. By adding relations (2. 1) and (2. 2), we obtain

$$\frac{1}{2}d(f+\bar{f}) = dg;$$

thus, g is equal to the real part of f with a real constant added it necessary. It remains to be proved that, if two holomorphic functions f_1 and f_2 in a neighbourhood of the same point have the same real part, then their difference $f = f_1 - f_2$ is constant. In fact, $d(f+\bar{f}) = 0$; that is,

$$\frac{\partial f}{\partial z}\,dz + \frac{\partial \bar{f}}{\partial \bar{z}}\,d\bar{z} = 0,$$

which implies that $\frac{\partial f}{\partial z} = 0$ and $\frac{\partial \bar{f}}{\partial \bar{z}} = 0$.

This completes the proof.

Note. Given a real harmonic function g in an open set D, there does not necessarily exist a holomorphic function f in the whole of D, whose real part is equal to g. For example, when D is the whole plane excluding the origin, $\log|z|$ is not the real part of a holomorphic function in D because the logarithm of z has no single-valued branch in D. Proposition 2. 2 says only that any real, harmonic function is *locally* the real part of a holomorphic function. However :

COROLLARY. *If* D *is a simply connected open set, any real harmonic function g is the real part of a holomorphic function f in* D.

For, the differential form $2\frac{\partial g}{\partial z}\,dz$ has a primitive in D (cf. chapter II, § 1, no. 7, theorem 3).

3. THE MEAN VALUE PROPERTY

We saw in chapter III, § 2, no. 1 that any holomorphic function f in an open set D has the mean value property : for any closed disc contained in D, the value of f at the centre of the disc is equal to the mean of its values on the boundary of the disc.

PROPOSITION 3. 1. *Any harmonic function in* D *has the mean value property.* It is sufficient to prove this for a real-valued harmonic function because the case of a complex-valued harmonic function reduces to the real case when the real and imaginary parts are consider separately.

Let g, then, be a real harmonic function in D and let S be a closed disc contained in D. By the corollary of proposition 2. 2, there exists a holomorphic function f in some neighbourhood of S whose real part is g. The value of f at the centre of S is equal to the mean of f on the boundary of the disc; by taking real parts, we see that the value of g at the centre of S is equal to its mean value on the boundary of the disc.
This completes the proof.

We shall see later (§ 4, no. 4) that, conversely, any continuous function with the mean value property is harmonic. In other words, the mean value property characterizes harmonic functions.
In chapter III, § 2, no. 2, we proved the *maximum modulus principle* for all (real-or complex valued) continuous functions with the mean value property. The maximum modulus principle therefore applies to harmonic functions.

4. Analyticity of harmonic functions

Proposition 4. 1. *Any harmonic function $g(x, y)$ in an open set D of the plane is an analytic function of the real variables x and y in D. In particular, any harmonic function is infinitely differentiable.*

Proof. We can suppose that g has real values and, as the proposition is local (since it is sufficient to show that g is analytic in a neighbourhood of each point of D), we shall suppose that $g(x, y)$ is harmonic in the open disc $x^2 + y^2 < \rho^2$. In this disc, g is the real part of a holomorphic function f, which can be expanded as a power series

$$f(z) = \sum_{n \geqslant 0} a_n z^n. \tag{4. 1}$$

Replace z by $x + iy$ in this series and consider the series

$$\sum_{n \geqslant 0} a_n (x + iy)^n \tag{4. 2}$$

as a power series in two variables x and y, it being understood, that $(x + iy)^n$ in (4. 2) is replaced by its expansion

$$(x + iy)^n = \sum_{p+q=n} \frac{n!}{p!\,q!} x^p (iy)^q. \tag{4. 3}$$

All the points (x,y) such that $|x| + |y| < \rho$ belong to the *domain of conver-*

gence of the double series (4. 2). For, if (x,y) is such a point, there exist $r_1 > |x|$ and $r_2 > |y|$ such that

$$r_1 + r_2 = r < \rho,$$

and we have

$$\sum_{p\geqslant 0,\, q\geqslant 0} \frac{(p+q)!}{p!\,q!} |a_{p+q}|(r_1)^p (r_2)^q = \sum_{n\geqslant 0} |a_n| r^n < +\infty.$$

In particular, the sum of the series (4. 2) is an *analytic* function in the product of the discs

(4. 4) $$|x| < \frac{\rho}{2}, \qquad |y| < \frac{\rho}{2}.$$

Let $\bar{f}(z) = \sum_{n\geqslant 0} \bar{a}_n z^n$ be the sum of the power series whose coefficients $\bar{a}_n$ are the complex conjugates of the coefficients of the series $f(z)$. Then,

(4. 5) $$2g(x, y) = f(x + iy) + \bar{f}(x - iy).$$

For the same reason as above, the function $\bar{f}(x - iy)$ is analytic in the open set (4. 4). Thus $g(x, y)$ is an analytic function in this open set. Hence the function g is analytic in some neighbourhood of the centre of any open disc in which it is harmonic. Proposition 4. 1 follows.

5. Method of finding a holomorphic function whose real part is given

We saw (proposition 2. 2) that any real harmonic function g is locally the real part of a holomorphic function f which is obtained by integration. We shall now see that, when g (which is analytic) is given by a power series expansion, f can be obtained without integrating.

Suppose again that $g(x, y)$ is harmonic in the open disc $x^2 + y^2 < \rho^2$ and revert to the notation of no. 4.

Consider the two formal power series in two variables X and Y :

$$f(X + iY) = \sum_{n\geqslant 0} a_n (X + iY)^n, \qquad \bar{f}(X - iY) = \sum_{n\geqslant 0} \bar{a}_n (X - iY)^n.$$

We have just seen that their domain of convergence contains the open set (4. 4). We now substitute for X and Y complex numbers x and y which satisfy (4. 4) and we obtain absolutely convergent series.

Let z be a complex number such that $|z| < \rho$. By (4. 5)

(5. 1) $$2g\left(\frac{z}{2}, \frac{z}{2i}\right) = f(z) + \bar{f}(0).$$

By putting z equal to 0 in this relation we obtain

$$2g(0, 0) = f(0) + \bar{f}(0).$$

Subtracting gives the formula

$$(5.2) \qquad 2g\left(\frac{z}{2}, \frac{z}{2i}\right) - g(0, 0) = f(z) + \frac{1}{2}(\bar{f}(0) - f(0)).$$

Thus the required function $f(z)$ *is equal, up to addition of a purely imaginary constant, to the known function*

$$(5.3) \qquad 2g\left(\frac{z}{2}, \frac{z}{2i}\right) - g(0, 0),$$

obtained by substituting complex variables in the double power series expansion of the function $g(x, y)$ *of the real variables* x *and* y.

Note. In the above calculation, we supposed the function $g(x, y)$ to be *harmonic* in the disc $x^2 + y^2 < \rho^2$. But relation (5. 2) still has a meaning for any real analytic function $g(x, y)$ with a power series expansion in the open set (4. 4); the function $f(z)$ which it defines has a power series expansion for $|z| < \rho$ and is therefore holomorphic in this disc. However, we can no longer be sure that g is the real part of the holomorphic function (5. 3). A suggested exercise is to show that a necessary and sufficient condition for g to be the real part of (5. 3) is that g is harmonic.

Example : consider the function

$$g(x, y) = \frac{\sin x \cos x}{\cos^2 x + \sinh^2 y}.$$

Then,

$$2g\left(\frac{z}{2}, \frac{z}{2i}\right) = \frac{2 \sin \frac{z}{2} \cos \frac{z}{2}}{\cos^2 \frac{z}{2} - \sin^2 \frac{z}{2}} = \tan z,$$

and, consequently,

$$f(z) = \tan z.$$

It can be verified that g is, in fact, the real part of $\tan z$; thus, the given function g is harmonic and is the real part of $\tan z$.

4. Poisson's Formula; Dirichlet's Problem

1. The integral representation of a harmonic function in a disc

Let $g(x, y)$ be a real harmonic function in the disc $x^2 + y^2 < \rho^2$; g is the real part of a holomorphic function

$$(1.1) \qquad f(z) = \sum_{n \geq 0} a_n z^n,$$

and we can suppose that a_0 is real, which determines the function f completely.

For $r < \rho$,

$$g(r\cos\theta, r\sin\theta) = a_0 + \frac{1}{2}\sum_{n\geqslant 1} r^n(a_n e^{in\theta} + \bar{a}_n e^{-in\theta}), \tag{1. 2}$$

the convergence being normal with respect to θ which varies from 0 to 2π. The right hand side of (1. 2) is a Fourier series expansion whose coefficients are given by the integral formulae

$$a_0 = \frac{1}{2\pi}\int_0^{2\pi} g(r\cos\theta, r\sin\theta)\, d\theta, \tag{1. 3}$$

$$a_n = \frac{1}{\pi}\int_0^{2\pi} \frac{g(r\cos\theta, r\sin\theta)}{(re^{i\theta})^n}\, d\theta \quad \text{for} \quad n \geqslant 1. \tag{1. 4}$$

Replace the coefficients a_n on the right hand side of (1. 1) by their values in (1. 3) and (1. 4). For $|z| < r$, we obtain

$$f(z) = \frac{1}{2\pi}\int_0^{2\pi} g(r\cos\theta, r\sin\theta)\left[1 + 2\sum_{n\geqslant 1}\left(\frac{z}{re^{i\theta}}\right)^n\right] d\theta \tag{1. 5}$$

since we can change the order of summation and integration because of the normal convergence. However,

$$1 + 2\sum_{n\geqslant 1}\left(\frac{z}{re^{i\theta}}\right)^n = \frac{re^{i\theta} + z}{re^{i\theta} - z},$$

from which we obtain the formula

$$f(z) = \frac{1}{2\pi}\int_0^{2\pi} g(r\cos\theta, r\sin\theta)\frac{re^{i\theta} + z}{re^{i\theta} - z}\, d\theta, \tag{1. 6}$$

which holds for $|z| < r$.

This integral formula expresses the holomorphic function $f(z)$ in the disc $|z| < r$ in terms of its real part on the boundary of the disc.

Let us equate the real parts of the two sides of (1. 6). We obtain

$$g(x, y) = \frac{1}{2\pi}\int_0^{2\pi} g(r\cos\theta, r\sin\theta)\frac{r^2 - |z|^2}{|re^{i\theta} - z|^2}\, d\theta \tag{1. 7}$$

(with $z = x + iy$).

This formula is true in the open disc $x^2 + y^2 < r^2$ for any real harmonic function g in the disc $x^2 + y^2 < \rho^2$ (with $r < \rho$). In fact, formula (1. 7) is also true for a *complex-valued* harmonic function g as one sees by separating the real and imaginary parts. Formula (1. 7) is called *Poisson's formula* and the function $\frac{r^2 - |z|^2}{|re^{i\theta} - z|^2}$, which occurrs in the integrand, is called *Poisson's kernel.*

2. Properties of Poisson's kernel

Fix r and θ; then, Poisson's kernel is a harmonic function of $z = x + iy$ defined at every point except $z = re^{i\theta}$. It is harmonic because it is the real part of the holomorphic function $\dfrac{re^{i\theta} + z}{re^{i\theta} - z}$. The Poisson kernel is zero at all points of the circle $|z| = r$ other than the point $z = re^{i\theta}$ and it is > 0 in the open disc $|z| < r$.

Let us now fix r and z with $|z| < r$. Then, the Poisson kernel is a periodic function of θ with strictly positive values; if we consider this function of θ as the density of a positive mass distribution on the unit circle, then the total mass of the distribution is equal to $+1$ because of the relation

$$\frac{1}{2\pi}\int_0^{2\pi} \frac{r^2 - |z|^2}{|re^{i\theta} - z|^2}\, d\theta = 1 \tag{2. 1}$$

which is deduced from (1. 7) putting g equal to the constant 1 (which is harmonic).

3. Dirichlet's problem for a disc

Dirichlet's problem is as follows : a continuous function is given on the circle of centre 0 and radius r by a continuous function $f(\theta)$ which is periodic of period 2π. We seek a function $F(z)$ of the complex variable z, which is defined and continuous in the closed disc $|z| \leqslant r$, which is harmonic in the open disc $|z| < r$ and which satisfies

$$F(re^{i\theta}) = f(\theta).$$

In other words we want to extend the given continuous function on the circle to a continuous function in the closed disc which is harmonic in the open disc. We shall confine ourselves to the case when both the given function f and the unknown function F are real-valued; the case of complex-valued functions can be reduced to the real-valued case by separating the real and imaginary parts.

Theorem. *The Dirichlet problem for a disc has a unique solution.*

First we prove the *uniqueness* of the solution if it exists. If F_1 and F_2 are two solutions to the problem, the difference $F_1 - F_2 = G$ is continuous in the closed disc, harmonic in the open disc and zero on the boundary of the disc. It is therefore sufficient to prove the following lemma :

Lemma. *A function* G, *which is defined and continuous in a closed disc, harmonic in the open disc and zero on the boundary of the disc, is identically zero.*

For, the closed disc is compact and so G attains its upper bound M at a point of the closed disc. If M were > 0, this point would be interior to the disc. By the maximum modulus principle (cf. chapter III, § 2), G would be constant and equal to M in the whole of the open disc and thus in the closed disc also because of continuity, which contradicts the hypothesis that G is zero on the boundary. For the same reason, the lower bound of G on the closed disc is 0. Thus G is identically zero.

The existence of a solution to Dirichlet's problem will be proved in the following section.

4. Solution of Dirichlet's problem for a disc

For $|z| < r$, let

$$F(z) = \frac{1}{2\pi}\int_0^{2\pi} f(\theta)\,\frac{r^2 - |z|^2}{|re^{i\theta} - z|^2}\,d\theta. \tag{4. 1}$$

We shall show that the function F so defined is harmonic and that

$$f(\theta_0) = \lim_{\substack{z \to re^{i\theta_0} \\ |z| < r}} F(z). \tag{4. 2}$$

Hence, the function F extended to the boundary of the disc by f is a solution of the Dirichlet problem, and this will complete the proof of the theorem in no. 3.

The function F defined in the interior of the disc by relation (4. 1) is obviously the real part of

$$\frac{1}{2\pi}\int_0^{2\pi} f(\theta)\,\frac{re^{i\theta} + z}{re^{i\theta} - z}\,d\theta,$$

and this is a holomorphic function of z in the open disc because it can be differentiated under the integration sign. Thus F is indeed harmonic in the open disc.

Relation (4. 2) remains to be proved. Here is the motivation of the proof: the Poisson kernel defines a positive mass distribution ε_z of total mass 1, which depends on the point z interior to the disc of radius r. We shall show that as z tends to a point $re^{i\theta_0}$ this mass distribution tends to the distribution which consists of a mass $+1$ situated at the point $re^{i\theta_0}$. A precise statement is that, given any arc $|\theta - \theta_0| \leqslant \eta$ of the circle of radius r containing the point $re^{i\theta_0}$, the total mass of the distribution ε_z *carried by this arc* tends to 1 when the point z tends to $re^{i\theta_0}$. An equivalent statement is that the total mass of the distribution ε_z carried by the complementary arc tends to 0 as z tends to the point $re^{i\theta_0}$ while remaining interior to the disc. Hence we want to prove the following:

LEMMA. *The integral*

$$(4.3)\qquad \frac{1}{2\pi}\int_{|\theta-\theta_0|>\eta}\frac{r^2-|z|^2}{|re^{i\theta}-z|^2}\,d\theta$$

tends to 0 *as* z *tends to* $re^{i\theta_0}$ *while its modulus remains* $< r$.

Proof of the lemma : put $z = \rho e^{i\alpha}$. If $|\alpha - \theta_0| \leqslant \frac{\eta}{2}$, then

$$|\alpha - \theta| \geqslant \frac{\eta}{2}$$

for all θ satisfying $|\theta - \theta_0| > \eta$. Thus we have under the integration sign

$$|re^{i\theta} - z| \geqslant r \sin\frac{\eta}{2},$$

and, consequently, the integral (4. 3) is bounded above by

$$\frac{1}{r^2\sin^2\frac{\eta}{2}}(r^2-\rho^2).$$

This indeed tends to 0 as ρ tends to r.
Having proved the lemma, we can now prove relation (4. 2). We have by (2. 1)

$$(4.4)\quad F(z) - f(\theta_0) = \frac{1}{2\pi}\int_{|\theta-\theta_0|\leqslant\eta}(f(\theta)-f(\theta_0))\frac{r^2-|z|^2}{|re^{i\theta}-z|^2}\,d\theta + \frac{1}{2\pi}\int_{|\theta-\theta_0|>\eta}(f(\theta)-f(\theta_0))\frac{r^2-|z|^2}{|re^{i\theta}-z|^2}\,d\theta.$$

Choose an $\varepsilon > 0$. The absolute value of the first integral of the right hand side of (4. 4) is bounded above by the upper bound of $|f(\theta) - f(\theta_0)|$ when $|\theta - \theta_0| \leqslant \eta$, since the total mass of the positive distribution ε_z is equal to 1. Since f is continuous, we can choose η so that the absolute value of the first integral is $\leqslant \frac{\varepsilon}{2}$. With this choice of η, we can give an upper bound $2Mm$ of the absolute value of the second integral on the right hand side of (4. 4), where M is an upper bound of $|f(\theta)|$ and m is the value of the integral (4. 3). By the above lemma, m tends to 0 as z tends to $re^{i\theta_0}$. Therefore, when z is sufficiently near to $re^{i\theta_0}$, the absolute value of the second integral will be $\leqslant \frac{\varepsilon}{2}$. Then,

$$|F(z) - f(\theta_0)| \leqslant \varepsilon,$$

which proves (4. 2).

The theorem of no. 3 is thus completely proved, and formula (4. 1) gives the solution of the Dirichlet problem explicitly.

5. CHARACTERIZATION OF HARMONIC FUNCTIONS BY THE MEAN VALUE PROPERTY

We have seen (§ 3, no. 3) that any harmonic function has the mean value property. The converse is also true :

THEOREM. *Any continuous function f in an open set D with the mean value property in D is harmonic in D.*

Proof. It is sufficient to show that f is harmonic in a neighbourhood of each point of D; to this end, we shall show that, if K is a closed disc contained in D, then f is harmonic in the interior of K. Consider the restriction of f to the boundary of the disc K; by the theorem of no. 3, there exists a continuous function F in K, which is harmonic in the interior of K and which coincides with f on the boundary of K. The difference $F - f$ is zero on the boundary of K and satisfies the maximum modulus principle in the interior of K because it has the mean value property. By the maximum modulus principle (cf. the lemma of no. 3), $F - f$ is identically zero in K. Thus f coincides with the harmonic function F in the interior of K and, consequently, f is indeed harmonic in the interior of K.

5. Holomorphic Functions of Several Complex Variables

1. DEFINITION OF A HOLOMORPHIC FUNCTION

Consider n complex variables $z_k = x_k + iy_k$ $(1 \leqslant k \leqslant n)$. By reasoning as in chapter II, § 2, no. 3, we see that the differential of a continuously differentiable function f can be written in the form

$$(1.1) \qquad df = \sum_{k=1}^{n} \left(\frac{\partial f}{\partial z_k} dz_k + \frac{\partial f}{\partial \bar{z}_k} d\bar{z}_k \right).$$

Keep all the variables fixed except z_k; a necessary and sufficient condition for the partial function to be a holomorphic function of z_k is that $\frac{\partial f}{\partial \bar{z}_k} = 0$. If this is so for each of the variables z_k, the differential df is a linear combination of the dz_k; conversely, if df is a linear combination of the dz_k, then the function f is holomorphic separately with respect to each variable z_k.

Definition. A function $f(z_1, \ldots, z_n)$ defined in an open set D of the space $\mathbf{C}^n$ of n variables z_k is said to be *holomorphic* in D if it is continuously differentiable and if, in addition, its differential df is equal to

$$\sum_k \frac{\partial f}{\partial z_k} dz_k.$$

It is clear that an analytic function of the complex variables z_k is holomorphic.

THEOREM. *A continuous function in an open set* D, *which is holomorphic separately with respect to each of the complex variable* z_k, *is holomorphic in* D *and also analytic in* D.

The proof of this theorem will be the main object of the next two sections. A particular consequence of the theorem is that any continuous function, which is holomorphic separately with respect to each variable z_k, is not only continuously differentiable but also infinitely differentiable. Another consequence is the equivalence of the concepts of holomorphy and analyticity for functions of several complex variables.

2. Cauchy's integral formula

First, we consider the case of two complex variables z_1 and z_2.

PROPOSITION 2. 1. *If* $f(z_1, z_2)$ *is continuous in the product of discs*

$$|z_1| < \rho_1, \qquad |z_2| < \rho_2 \tag{2. 1}$$

and holomorphic separately with respect to z_1 *and* z_2 *in* (2. 1), *then, when*

$$|z_k| < r_k < \rho_k \quad (k = 1, 2),$$

we have

$$f(z_1, z_2) = \frac{1}{(2\pi i)^2} \iint \frac{f(\zeta_1, \zeta_2)\, d\zeta_1\, d\zeta_2}{(\zeta_1 - z_1)(\zeta_2 - z_2)}, \tag{2. 2}$$

where the double integral is taken over the product of the circles $|\zeta_1| = r_1$ *and* $|\zeta_2| = r_2$, *each being described in the positive sense.*

Proof. Fix z_2 in the open disc $|z_2| < r_2$. The function $f(z_1, z_2)$ is holomorphic in z_1 in the disc $|z_1| < \rho_1$. We can thus apply Cauchy's integral formula (chapter II, § 2, no. 5) to it, giving

$$f(z_1, z_2) = \frac{1}{2\pi i} \int_{|\zeta_1| = r_1} \frac{f(\zeta_1, z_2)}{\zeta_1 - z_1} d\zeta_1 \quad \text{for} \quad |z_1| < r_1. \tag{2. 3}$$

Now fix ζ_1 such that $|\zeta_1| = r_1$. The function $f(\zeta_1, z_2)$ is holomorphic in z_2 for $|z_2| < \rho_2$, so, similarly, we obtain

$$f(\zeta_1, z_2) = \frac{1}{2\pi i} \int_{|\zeta_2| = r_2} \frac{f(\zeta_1, \zeta_2)}{\zeta_2 - z_2} d\zeta_2 \quad \text{for} \quad |z_2| < r_2. \tag{2. 4}$$

Substitute the value of $f(\zeta_1, z_2)$ from (2. 4) in the integrand of (2. 3). Since the function $f(\zeta_1, \zeta_2)$ is continuous, we obtain formula (2. 2) exactly.

Note (due to Hartogs). Let $f(z_1, z_2)$ be a continuous function defined in the union of the two open sets (with very small $\varepsilon > 0$)

$$\text{(A)} \qquad |z_1| < \rho_1, \qquad |z_2| < \varepsilon,$$

$$\text{(B)} \qquad \rho_1 - \varepsilon < |z_1| < \rho_1, \qquad |z_2| < \rho_2.$$

We suppose that, in (A), f is a holomorphic function of z_1 and that, in (B), f is a holomorphic function of z_2. Then f can be extended to a holomorphic function of the variables z_1 and z_2 in the open set (2. 1) and the extended function satisfies the integral formula (2. 2).

Indication of proof. Choose r_1 and r_2 arbitrarily so that $r_1 < \rho_1$ and $r_2 < \rho_2$, but large enough for $\varepsilon < r_2$, $r > \rho_1 - \varepsilon$. We shall show that f can be extended to another function, again denoted by $f(z_1, z_2)$, which is holomorphic in the open set

$$(2.\ 5) \qquad |z_1| < r_1, \qquad |z_2| < r_2,$$

and which satisfies (2. 2) in this open set. First, relation (2. 3) holds for $|z_1| < r_1$, $|z_2| < \varepsilon$ because f is a holomorphic function of z_1 in (A); next, if $|\zeta_1| = r_1$, relation (2. 4) holds for $|z_2| < r_2$ because f is a holomorphic function of z_2 in (B). Thus (2. 2) holds for $|z_1| < r_1$, $|z_2| < \varepsilon$. However, the right hand side of (2. 2) is a holomorphic function of z_1 and z_2 in (2. 5); if we write $f(z_1, z_2)$ for the function thus extended, it satisfies (2. 2) in (2. 5). This completes the proof.

Proposition 2. 1 has an analogue for functions of n complex variables. In this case, the integral formula (2. 1) is replaced by

$$f(z_1, \dots, z_n) = \left(\frac{1}{2\pi i}\right)^n \int \cdots \int \frac{f(\zeta_1, \dots, \zeta_n)\,d\zeta_1 \dots d\zeta_n}{(\zeta_1 - z_1)\dots(\zeta_n - z_n)}.$$

3. Series expansion of a holomorphic function

Proposition 3. 1 *Under the hypotheses of proposition* 2. 1, *the function* f *has a double power series expansion in the open set* (2. 1)

$$(3.\ 1) \qquad f(z_1, z_2) = \sum_{p,\, q \geqslant 0} a_{p,q}(z_1)^p (z_2)^q.$$

The proof will be similar to that given in the case of one complex variable (cf. chapter II, § 2, no. 6, theorem 3). We know already that, if the power series expansion exists, it is *unique* because it is necessarily the Taylor expansion of f at the origin It is sufficient then, given r_1' and r_2' such that

$$r_1' < \rho_1, \qquad r_2' < \rho_2,$$

to find a double power series which converges normally to $f(z_1, z_2)$ in the product of the discs

$$|z_1| \leqslant r_1', \qquad |z_2| \leqslant r_2'.$$

We choose r_1 and r_2 such that $r_1' < r_1 < \rho_1$, $r_2' < r_2 < \rho_2$ and apply the integral formula (2. 2) for $|z| \leqslant r_1'$, $|z_2| \leqslant r_2'$. Now,

$$\text{(3. 2)} \qquad \frac{1}{(\zeta_1 - z_1)(\zeta_2 - z_2)} = \sum_{p,\, q \geqslant 0} \frac{(z_1)^p (z_2)^q}{(\zeta_1)^{p+1}(\zeta_2)^{q+1}},$$

and this series is normally convergent for $|z_i| \leqslant r_i'$, $|\zeta_i| = r_i$ $(i = 1, 2)$. We substitute this value of $\dfrac{1}{(\zeta_1 - z_1)(\zeta_2 - z_2)}$ from (3. 2) into the integrand on the right hand side of (2. 2). Because of the normal convergence, we can integrate term by term and we obtain (3. 1) exactly, where the coefficients $a_{p,q}$ are given by the integral formula

$$\text{(3. 3)} \qquad a_{p,q} = \frac{1}{(2\pi i)^2} \iint \frac{f(\zeta_1, \zeta_2)}{(\zeta_1)^{p+1}(\zeta_2)^{q+1}} d\zeta_1 \, d\zeta_2.$$

Proposition 3. 1 is thus proved.
There is an analogous proposition for n complex variables.

It is clear that the theorem stated at the end of no. 1 follows from proposition 3. 1.

Note. It can be shown that a function $f(z_1, \ldots, z_n)$, which is holomorphic separately with respect to each variable in an open set D, is *continuous* in D, and, consequently, holomorphic. The proof is difficult and will not be given here.

4. Calculation of the coefficients of the Taylor expansion of a holomorphic function

As in the case of one variable, the coefficients $a_{p,q}$ can be expressed as integrals involving the function f. To do so, we replace z_1 by $r_1 e^{i\theta_1}$ and z_2 by $r_2 e^{i\theta_2}$ in relation (3. 1); integrating term by term gives

$$\text{(4. 1)} \qquad a_{p,q}(r_1)^p (r_2)^q = \frac{1}{4\pi^2} \int_0^{2\pi} \int_0^{2\pi} f(r_1 e^{i\theta_1}, r_2 e^{i\theta_2}) e^{-i(p\theta_1 + q\theta_2)} d\theta_1 d\theta_2.$$

We deduce the Cauchy inequalities

$$\text{(4. 2)} \qquad |a_{p,q}| \leqslant \frac{M(r_1, r_2)}{(r_1)^p (r_2)^q},$$

where $M(r_1, r_2)$ denotes the upper bound of $|f(z_1, z_2)|$ for $|z_1| = r_1$, $|z_2| = r_2$, or, which is the same, for $|z_1| \leqslant r_1$, $|z_2| \leqslant r_2$. The reader is left to state the theorem similar to Liouville's theorem and also the maximum modulus principle.

5. Composition of holomorphic functions

Proposition 5. 1. *Let $f(z_1, \ldots, z_n)$ be a holomorphic function in an open set D of $\mathbf{C}^n$ and let*

$$g_1, \ldots, g_n$$

be holomorphic functions in an open set D′ of $\mathbf{C}^p$, whose values at each point of D′ are coordinates of a point in D. The function $f \circ g$ defined by

$$(t_1, \ldots, t_p) \to f(g_1(t_1, \ldots, t_p), g_2(t_1, \ldots, t_p), \ldots, g_n(t_1, \ldots, t_p))$$

is a holomorphic function of $t_1, \ldots, t_p$ in the open set D′.

Proof. We could use substitution in convergent power series but, since we have not gone into the details of this matter in the case of several variables, we shall give a method based on an entirely different principle.

Since f is holomorphic, we have, by hypothesis,

$$df = \sum_{k=1}^{n} \frac{\partial f}{\partial z_k} dz_k, \tag{5. 1}$$

and, since the functions g_k are holomorphic,

$$dz_k = \sum_{j=1}^{p} \frac{\partial g_k}{\partial t_j} dt_j. \tag{5. 2}$$

We find the differential of the composed function $f \circ g$ by substituting for the differentials dz_k in (5. 1), their values in (5. 2); thus $d(f \circ g)$ is a linear combination of the dt_j, and, consequently, $f \circ g$ is a holomorphic function of the t_j.

6. The implicit function theorem

Proposition 6. 1. *Let $f_j(x_1, \ldots, x_n; z_1, \ldots, z_p)$, for $j = 1, \ldots, n$, be holomorphic functions in a neighbourhood of a point $x_j = a_j$, $z_k = c_k$, and suppose that the functional determinant* $\det\left(\frac{\partial f_j}{\partial x_{j'}}\right)$ *is $\neq 0$ at the point considered. Then, the equations*

$$y_j = f_j(x_1, \ldots, x_n; z_1, \ldots, z_p) \quad (j = 1, \ldots, n) \tag{6. 1}$$

can be solved when the x_j *are sufficiently near to the* a_j, *when the* z_k *are sufficiently near to the* c_k *and when the* y_j *are sufficiently near to the* $b_j = f_j(a_1, \ldots, a_n; c_1, \ldots, c_p)$, *in the following way* :

$$x_j = g_j(y_1, \ldots, y_n; z_i, \ldots, z_p), \tag{6. 2}$$

where the g_j *are holomorphic functions in some neighbourhood of the point* $(b_1, \ldots, b_n; c_1, \ldots, c_p)$.

Proof. We shall reduce it to the classical implicit function theorem for real variables. Put

$$x_j = x'_j + ix''_j, \qquad y_j = y'_j + iy''_j,$$

x'_j, x''_j, y'_j and y''_j being real. The exterior product $dx_j \wedge d\bar{x}_j$ is equal to

$$(dx'_j + i\,dx''_j) \wedge (dx'_j - i\,dx''_j) = -2i\,dx'_j \wedge dx''_j.$$

Hence,

$$dx_j \wedge d\bar{x}_j = -2i\,dx'_j \wedge dx''_j, \quad \text{and, similary,} \quad dy_j \wedge d\bar{y}_j = -2i\,dy'_j \wedge dy''_j. \tag{6. 3}$$

When $z_1, \ldots, z_p$ are kept fixed,

$$dy_1 \wedge \cdots \wedge dy_n = \det\left(\frac{\partial f_j}{\partial x_{j'}}\right) dx_1 \wedge \cdots \wedge dx_n,$$

$$d\bar{y}_1 \wedge \cdots \wedge d\bar{y}_n = \det\left(\frac{\partial \bar{f}_j}{\partial \bar{x}_{j'}}\right) d\bar{x}_1 \wedge \cdots \wedge d\bar{x}_n,$$

whence, by multiplication,

$$dy_1 \wedge d\bar{y}_1 \wedge \cdots \wedge dy_n \wedge d\bar{y}_n = \left|\det\left(\frac{\partial f_j}{\partial x_{j'}}\right)\right|^2 dx_1 \wedge d\bar{x}_1 \wedge \cdots \wedge dx_n \wedge d\bar{x}_n.$$

This relation and (6. 3) show that the functional determinant of $y'_1, y''_1, \ldots, y'_n, y''_n$ with respect to $x'_1, x''_1, \ldots, x'_n, x''_n$ is equal to

$$\left|\det\left(\frac{\partial f_j}{\partial x_{j'}}\right)\right|^2,$$

which is $\neq 0$ at the point $(a_1, \ldots, a_n; c_1, \ldots, c_p)$ by hypothesis. We now apply the implicit function theorem : $x'_1, x''_1, \ldots, x''_n$ can be expressed (locally) as continuously differentiable functions of $y'_1, y''_1, \ldots, y''_n$ and the real and imaginary parts of $z_1, \ldots, z_p$. However, the system of linear equations

$$dy_j = \sum_{j'} \frac{\partial f_j}{\partial x_{j'}} dx_{j'} + \sum_k \frac{\partial f_j}{\partial z_k} dz_k$$

shows that the $dx_{j'}$ are linear combinations of the dy_j and the dz_k. Thus, $x_1, \ldots, x_n$ are, in fact, *holomorphic* functions of the y_i and the z_k.
This completes the proof.

Exercises

1. Show that, if $f(z)$ is holomorphic in an open set D, then, for all $z \in D$,

(i) $$\Delta |f(z)|^2 = 4|f'(z)|^2,$$

(ii) $$\Delta \log(1 + |f(z)|^2) = 4|f'(z)|^2/(1 + |f(z)|^2)^2,$$

where Δ denotes the Laplacian defined in § 3, no. 1.

2. (i) Let $g(z)$ be a holomorphic function in the disc $|z| < R$. Show that, if $0 \leqslant r < R$ and if $g(z)$ has no zeros in the closed disc $|z| \leqslant r$, then

$$\log|g(0)| = \frac{1}{2\pi}\int_0^{2\pi} \log|g(re^{i\theta})|\, d\theta.$$

(ii) Show that the integral

$$\int_0^{2\pi} \log|re^{i\theta} - re^{it}|\, d\theta$$

exists and that its value is equal to $2\pi \log r$ (r, t real, $r > 0$). Deduce that, if $f(z) \neq 0$ is a meromorphic function in the disc $|z| < R$, and if $0 < r < R$, the integral $\int_0^{2\pi} \log|f(re^{i\theta})|\, d\theta$ is convergent.

(iii) Let $a_1, a_2, \ldots, a_p$ be the zeros and $b_1, b_2, \ldots, b_q$ the poles of the function $f(z)$ considered in (ii) which are contained in the punctured disc $0 < |z| \leqslant r$ (each being counted as many times as its order of multiplicity), and let

$$f(z) = c_n z^n + c_{n+1} z^{n+1} + \cdots$$

be the Laurent expansion of f at the origin $\left(\text{so } n \text{ is an integer} \gtreqless 0\right)$. Show that

$$\frac{1}{2\pi}\int_0^{2\pi} \log|f(re^{i\theta})|\, d\theta = \log|c_n| - \sum_{j=1}^{p} \log|a_j| + \sum_{k=1}^{q} \log|b_k| + (n + p - q)\log r.$$

(Consider the function

$$g(z) = f(z)\left(\frac{r}{z}\right)^n \prod_{1 \leqslant j \leqslant p} \frac{r^2 - \bar{a}_j z}{r(z - a_j)} \prod_{1 \leqslant k \leqslant q} \frac{r(z - b_k)}{r^2 - \bar{b}_k z},$$

and show that it is holomorphic without zeros in an open set containing the closed disc $|z| \leqslant r$ and that $|g(z)| = |f(z)|$ if $|z| = r$.)

3. The harmonic functions considered in this question are all assumed to be real-valued.

(i) If $f(z)$ is harmonic in the disc $|z| < R$ and if $f(z) \geqslant 0$ throughout the disc, prove the inequalities

$$\frac{R - |z|}{R + |z|} f(0) \leqslant f(z) \leqslant \frac{R + |z|}{R - |z|} f(0)$$

for all $|z| < R$. (Use the Poisson formula and note that the Poisson kernel satisfies the inequalities

$$\frac{r - |z|}{r + |z|} \leqslant \frac{r^2 - |z|^2}{|re^{i\theta} - z|^2} \leqslant \frac{r + |z|}{r - |z|} \quad (\text{for } |z| < r).)$$

(ii) Deduce from (i) that, if $f(z)$ is harmonic and $\geqslant 0$ in the disc $D(a, r)$ of centre a and radius r, then

$$\frac{1}{3} f(a) \leqslant f(z) \leqslant 3 f(a)$$

for all z in the disc $D(a, r/2)$.

(iii) Let $f(z)$ be a harmonic function $\geqslant 0$ in a connected open set of the plane $\mathbf{C}$ and let K be a compact subset of D. Show that there is a constant M, which depends only on D and K, such that

$$f(z_1) \leqslant M f(z_2)$$

for all z_1, z_2 in K. (Show that there is a finite number of closed discs D_n satisfying the conditions that

$$D \supset \bigcup_n D_n \supset K,$$

and that, for any pair of them, D_p' and D_q, say, there exists a sequence $D_{n_1}, \ldots, D_{n_k}$ such that $D_{n_1} = D_p$, $D_{n_k} = D_q$ and $D_{n_{j-1}} \cap D_{n_j} \neq \emptyset$ for $j = 2, 3, \ldots, k$. Apply (ii) to each of these discs.)

(iv) Let $(f_n(z))$ be a sequence of harmonic functions in a connected open set D and let the sequence be monotonic increasing (in the broad sense), that is

$$f_n(z) \leqslant f_{n+1}(z) \qquad \text{for all } z \in D \text{ and } n = 1, 2, \ldots.$$

If there exists an $a \in D$ such that $\sup_n |f_n(a)| < \infty$, show that the sequence $(f_n(z))$ converges uniformly in any compact subset in D to a harmonic function (note the equivalence of the convergence of the sequence $(f_n(z))$ and of the series $\Sigma(f_{n+1}(z) - f_n(z))$ and apply (iii).)

4. *Subharmonic functions.* A real-valued continuous function in an open set D of the plane $\mathbf{C}$ is said to be *subharmonic* if, for any $a \in D$,

(SH) $$f(a) \leqslant \frac{1}{2\pi} \int_0^{2\pi} f(a + re^{i\theta})\, d\theta$$

for sufficiently small $r > 0$.

(i) If $f(z)$ is holomorphic in an open set D, show that $|f(z)|^p$ is subharmonic in D for $p \geqslant 0$.

(ii) If $f_\nu(z)$, $\nu = 1, 2, \ldots, n$, are subharmonic in D, then the following functions are also subharmonic in D:

$$\sum_{\nu=1}^{n} a_\nu f_\nu(z), \quad a_\nu \geqslant 0; \qquad \sup_{1 \leqslant \nu \leqslant n} f_n(z).$$

(iii) If a sequence of subharmonic functions $f_n(z)$ in D converges uniformly on any compact subset of D, then the limit function is also subharmonic.

(iv) Show that the maximum modulus principle holds for subharmonic functions; show, in fact, that:

(1) If f is a subharmonic function in an open set D and if f has a relative maximum at a point $a \in D$ (i.e. $f(z) \leqslant f(a)$ for any z sufficiently near to a), then f is constant in some neighbourhood of a.

(2) Let D be a bounded, connected, open subset of the plane, let f be a continuous function in $\bar{D}$ which is subharmonic in D and let M be the upper bound of $f(z)$ as z describes the frontier of D. Then,

(*a*) $$f(z) \leqslant M \text{ for all } z \in D,$$

(*b*) if $f(a) = M$ at a point $a \in D$, then f is constant.

(v) Let Γ be the oriented boundary of a compact subset K of an open set D. Show that, if u, v are two (real-valued) functions with continuous second derivatives, then

$$\iint_K (v\Delta u - u\Delta v)\, dx\, dy = \int_\Gamma \left(u\frac{\partial v}{\partial y} - v\frac{\partial u}{\partial y}\right) dx + \left(v\frac{\partial u}{\partial x} - u\frac{\partial v}{\partial x}\right) dy.$$

(Use the Green-Riemann formula quoted in chapter II, § 1, no. 9; first take $P = -v\dfrac{\partial u}{\partial y}$, $Q = v\dfrac{\partial u}{\partial x}$, then interchange u and v.) Deduce that, if $f(z)$ is a function defined in D with continuous second derivatives and if $a \in D$, then

$$\iint_{|z-a| \leqslant r} (\Delta f)(z)\, dx dy = \int_0^{2\pi} \frac{\partial f}{\partial r}(a + re^{i\theta})\, r\, d\theta,$$

for sufficiently small $r > 0$. (Put $u = f$ and $v = 1$ in the previous relation). Deduce that

$$\frac{1}{2\pi}\int_0^{2\pi} f(a + \rho e^{i\theta})\, d\theta = f(a) + \int_0^{\rho} \frac{dr}{2\pi r} \iint_{|z-a|\leqslant r} (\Delta f)(z)\, dxdy$$

for sufficiently small $\rho > 0$ and show that a function $f(z)$ with continuous second derivatives is subharmonic in D if and only if $\Delta f(z) > 0$ for $z \in D$.

Example. Show that, if $f(z)$ is holomorphic in an open set D, the function $\log(1 + |f(z)|^2)$ is subharmonic in D.

5. Let $f(z)$ be a subharmonic function in the disc $|z| < R$. Show that, if $0 < r_1 < R$ and if $g(z)$ is the solution to the Dirichlet problem in the disc $|z| \leqslant r_1$ such that $g(r_1 e^{i\theta}) = f(r_1 e^{i\theta})$, then

$$g(re^{i\theta}) \geqslant f(re^{i\theta})$$

for $0 \leqslant r < r_1$. Deduce that the function

$$m(r) = \frac{1}{2\pi}\int_0^{2\pi} f(re^{i\theta})\, d\theta$$

is a continuous, monotonic increasing (in the broad sense) function of r in $0 \leqslant r < R$.

6. Show that, if $f(z)$ is holomorphic in the disc $|z| < R$ and α is a real number > 0, then

$$I_\alpha(r) = \frac{1}{2\pi}\int_0^{2\pi} |f(re^{i\theta})|^\alpha\, d\theta$$

is continuous and monotonic increasing (in the broad sense) in $0 \leqslant r < R$

CHAPTER V

Convergence of Sequences of Holomorphic or Meromorphic Functions; Series, Infinite Products; Normal Families

In this chapter we shall only consider functions of one complex variable. However, many of the concepts can be extended to the case of several complex variables.

1. Topology of the Space $\mathcal{C}(D)$

1. Uniform convergence on compact subsets

Let D be an open set of the complex plane **C**. We shall write $\mathcal{C}(D)$ for the vector space of *continuous* (complex-valued) functions in the open set D. Similarly, $\mathcal{H}(D)$ denotes the vector space of *holomorphic* functions in D.

Definition. We say that a sequence of functions $f_n \in \mathcal{C}(D)$ *converges uniformly on compact subsets* if, for any compact subset $K \subset D$, the sequence of restrictions $f_n|K$ is uniformly convergent.

This definition applies in the particular case of functions of the space $\mathcal{H}(D)$.

We know that the limit of a uniformly convergent sequence of continuous functions is a continuous function. Thus, if the sequence of continuous functions f_n is uniformly convergent on any compact subset of D, the limit function f is such that its rectriction $f|K$ to any compact subset $K \subset D$ is continuous. Since any point of D has a compact neighbourhood contained in D, it follows that f is continuous.

Definition. We say that a series $\sum_n f_n$ of functions $f_n \in \mathcal{C}(D)$ *converges normally in compact subsets* of D if, for any compact subset $K \subset D$, the series of restrictions $f_n|K$ converges normally. In other words, on any compact subset K, the given series is majorized by a convergent series of constant positive terms. It is clear that, if a series is normally convergent on compact subsets, the partial sums of this series form a sequence which is uniformly convergent on compact subsets.

PROPOSITION 1.1 *A sufficient (and necessary) condition for a sequence of functions $f_n \in \mathcal{C}(D)$ to be uniformly convergent on compact subsets of D is that the sequence of restrictions $f_n|\Sigma$ to every compact disc $\Sigma \subset D$ is uniformly convergent. There is a similar theorem for normally convergent series.*

For, any compact subset K of D can be covered by the interiors of a finite number of compact discs contained in D. The proposition follows immediately.

2. FUNDAMENTAL THEOREMS ON THE CONVERGENCE OF HOLOMORPHIC FUNCTIONS

THEOREM 1. *If a sequence of functions $f_n \in \mathcal{H}(D)$ is uniformly convergent on compact subsets, then the limit function f is holomorphic in D.*

Proof. We have just seen that f is continuous in D. To show that f is holomorphic, it is sufficient, by Morera's theorem (chapter II, § 2, no. 7, theorem 4) to show that the differential form $f(z)\,dz$ is closed. For this, it is sufficient to show that $\int_\gamma f(z)\,dz = 0$ whenever γ is the boundary of a rectangle contained in D (cf. chapter II, § 1, proposition 4.1). However, f is the uniform limit of the sequence f_n on the boundary of each rectangle, and, hence,

$$\int_\gamma f(z)\,dz = \lim_n \int_\gamma f_n(z)\,dz = 0,$$

which proves theorem 1.

CAROLLARY. *The sum of a series of holomorphic functions, which is normally convergent on compact subsets of D, is a holomorphic function in D.*

THEOREM 2. *If a sequence of functions $f_n \in \mathcal{H}(D)$ converges to $f \in \mathcal{H}(D)$ uniformly on compact subsets, then the sequence of derivatives f_n' converge to the derivative f' uniformly on compact subsets.*

Proof. By proposition 1.1, it is sufficient to show that the f_n' converge to f' uniformly on any compact disc in D. Let Σ be such a disc of radius r and choose the centre of Σ as origin 0. There exists $r_0 > r$ such that the closed disc of centre 0 and radius r_0 is contained in D. Thus the f_n are holomorphic for $|z| < r_0 + \varepsilon$ (for sufficiently small $\varepsilon > 0$) and converge to f uniformly for $|z| \leqslant r_0$.

We shall show that the derivatives f'_n converge uniformly to f' for $|z| \leqslant r$, which will follow immediately from the following lemma :

LEMMA. *If* $g(z)$ *is holomorphic for* $|z| < r_0 + \varepsilon$ *and if* $|g(z)| \leqslant M$ *for* $|z| \leqslant r_0$, *then*

$$|g'(z)| \leqslant M \frac{r_0}{(r_0 - r)^2} \quad \text{for} \quad |z| \leqslant r < r_0. \tag{2. 1}$$

Proof of the lemma. There is a convergent expansion

$$g(z) = \sum_{n \geqslant 0} a_n z^n \qquad \text{for} \quad |z| \leqslant r_0. \tag{2. 2}$$

By Cauchy's inequalities, we have $|a_n| \leqslant \frac{M}{(r_0)^n}$. On the other hand, differentiating term by term gives

$$g'(z) = \sum_{n \geqslant 0} n a_n z^{n-1}. \tag{2. 3}$$

Thus, for $|z| \leqslant r < r_0$,

$$|g'(z)| \leqslant \frac{M}{r_0} \sum_{n \geqslant 0} \frac{n r^{n-1}}{(r_0)^{n-1}}. \tag{2. 4}$$

We shall find the sum of the series $\sum_{n \geqslant 0} n \left(\frac{r}{r_0}\right)^{n-1}$; since nt^{n-1} is the derivative of t^n, then $\sum_n nt^{n-1}$ is the derivative of $\sum t^n = \frac{1}{1-t}$, and, consequently,

$$\sum_{n \geqslant 0} n \left(\frac{r}{r_0}\right)^{n-1} = \frac{1}{\left(1 - \frac{r}{r_0}\right)^2}.$$

This substituted in (2. 4) gives the inequality

$$|g'(z)| \leqslant \frac{M}{r_0} \frac{1}{\left(1 - \frac{r}{r_0}\right)^2},$$

which proves the lemma.

Note. One can construct another proof of theorem 2 by observing that Cauchy's integral formula

$$f(z) = \frac{1}{2\pi i} \int_\gamma \frac{f(t)\, dt}{t - z}$$

(where γ denotes the boundary of a disc concentric with Σ but with a slightly larger radius) gives, by differentiating with respect to z under the integral sign,

$$f'(z) = \frac{1}{2\pi i} \int_\gamma \frac{f(t)\, dt}{(t - z)^2}.$$

Thus,

$$f'(z) = \lim_n \frac{1}{2\pi i}\int_\gamma \frac{f_n(t)\,dt}{(t-z)^2} = \lim_n f'_n(z).$$

and the limit is approached uniformly for $z \in \Sigma$.

PROPOSITION 2. 1. *Let D be a connected open set. If a sequence of holomorphic functions $f_n \in \mathcal{H}(D)$ is uniformly convergent on compact subsets of D, and if each f_n is $\neq 0$ at any point of D, then the limit function f is $\neq 0$ at any point of D unless it is identically zero.*

Proof. Suppose f is not identically zero. Then the zeros of f (which is holomorphic by theorem 1) are *isolated* since D is connected. Suppose f vanishes at z_0; by proposition 4. 1 of chapter III, § 5, the order of multiplicity of this zero is equal to the integral

$$\frac{1}{2\pi i}\int_\gamma \frac{f'(z)\,dz}{f(z)},$$

taken round a circle γ of small radius centred at z_0. By theorem 2, this integral is the limit of integrals

$$\frac{1}{2\pi i}\int_\gamma \frac{f'_n(z)\,dz}{f_n(z)},$$

and these integrals are zero since the holomorphic functions f_n do not vanish. Hence, we have a contradiction and this proves the proposition.

Definition. A function defined in an open set D is said to be *simple* if the mapping it defines is injective, in other words, if it always takes distinct values at distinct points.

PROPOSITION 2. 2. *Let D be an open set of* **C**. *If a sequence of holomorphic functions $f_n \in \mathcal{H}(D)$ converges uniformly on compact subsets of D and if each f_n is simple, then the limit function f is simple if it is not constant.*

Proof. We use *reductio ad absurdum.* Let us assume that $f(z_1) = f(z_2) = a$ for two distinct points z_1 and z_2 of D and we shall show that this leads to a contradiction. Consider two open discs S_1 and S_2 with centres z_1 and z_2 an with radii so small that S_1 and S_2 are disjoint and are contained in D. By proposition 2. 1, f_n takes the value a in S_1 and in S_2 for sufficiently large n, which contradicts the simplicity of f_n.

3. TOPOLOGY OF THE SPACE $\mathcal{C}(D)$

We have already defined what we mean by a sequence of functions $f_n \in \mathcal{C}(D)$ which *converges uniformly on compact subsets.* We shall now define a *topology* on the vector space $\mathcal{C}(D)$ in a more precise way. The vector subspace $\mathcal{H}(D)$ is given the induced topology. For any pair (K, ε) consisting of a compact subset $K \subset D$ and a number $\varepsilon > 0$, we consider the subset $V(K, \varepsilon)$ of $\mathcal{C}(D)$ defined by

$$f \in V(K, \varepsilon) \iff |f(x)| \leqslant \varepsilon \quad \text{for} \quad x \in K. \tag{3.1}$$

A necessary and sufficient condition for a sequence of functions $f_n \in \mathcal{C}(D)$ to converge to f uniformly on compact subsets is that, for any K and ε,

$$f - f_n \in V(K, \varepsilon) \quad \text{for sufficiently large } n.$$

This expresses that the sequence of $f_n \in \mathcal{C}(D)$ has f as its *limit* in the topology (if it exists) for which the sets $V(K, \varepsilon)$ form a fundamental system of neighbourhoods of 0 (the neighbourhoods of a point f are defined by translating the neighbourhoods of 0 by f).

PROPOSITION 3. 1. *$\mathcal{C}(D)$ has indeed a topology (invariant under translation) in which the sets $V(K, \varepsilon)$ form a fundamental system of neighbourhoods of 0. This topology is unique and can be defined by a distance which is invariant under translation.*

Proof. The uniqueness of the topology is obvious, because we know a fundamental system of neighbourhoods of 0, and, therefore, a fundamental system of neighbourhoods of any point of the space $\mathcal{C}(D)$ by translation. We need only find a distance function, which is invariant under translation, such that the $V(K, \varepsilon)$ form a fundamental system of neighbourhoods of 0 in the topology defined by this distance.

First we introduce the concept of an *exhaustive sequence of compact subsets* of D, that is, an increasing sequence of compact sets $K_i \subset D$ (thus $K_i \subset K_{i+1}$) such that any compact subset K contained in D is contained in one of the K_i.

LEMMA. *There exists an exhaustive sequence of compact subsets of* D.

For, consider the compact discs contained in D whose centre has rational coordinates and whose radius is rational. They form a *countable* set that can be arranged in a sequence $D_1, D_2, \ldots, D_n, \ldots$. Let

$$K_i = \bigcup_{n \leqslant i} D_n.$$

We now show that the K_i form an exhaustive sequence : the interiors of the discs D_n form an open cover of D and, consequently, any compact subset K of D is contained in a K_i.

Suppose, from now on, that we have chosen an exhaustive sequence of compact subsets K_i and, for each $f \in \mathcal{C}(D)$, put

$$M_i(f) = \sup_{z \in K_i} |f(z)|, \tag{3.2}$$

$$d(f) = \sum_{i \geq 1} 2^{-i} \inf(1, M_i(f)). \tag{3.3}$$

We note that $d(f)$ is finite because the series on the right hand side is majorized by the geometric series $\sum_{n \geq 1} 2^{-i}$. We shall prove that $d(f)$ has the following properties:

$$d(f) = 0 \Longleftrightarrow f = 0, \tag{3.4}$$

$$d(f+g) \leqslant d(f) + d(g), \tag{3.5}$$

$$\begin{cases} 2^{-i} \inf(1, M_i(f)) \leqslant d(f), \\ d(f) \leqslant M_i(f) + 2^{-i}. \end{cases} \tag{3.6}$$

Proof of (3.4). It is clear that, if f is identically zero, $d(f) = 0$; conversely, $d(f) = 0$ implies, by (3.3), that $M_i(f) = 0$ for all i, so the restriction of to each open set K_i is zero, and, consequently, f is identically zero.

Proof of (3.5). It is obvious that

$$M_i(f+g) \leqslant M_i(f) + M_i(g),$$

from which we easily deduce that

$$\inf(1, M_i(f+g)) \leqslant \inf(1, M_i(f)) + \inf(1, M_i(g)),$$

and (3.5) follows by summation.

The relations (3.4) and (3.5) show that, if the distance between f and g is defined to be equal to $d(f-g)$, this distance function is a *metric* satisfying the triangle inequality; this metric is *invariant under translation*. It defines a Hausdorff topology on the space $\mathcal{C}(D)$ which is invariant under translation.

We now prove the inequalities (3.6). The first follows obviously from the definition (3.3). On the other hand, if i is an integer $\geqslant 1$, then

$$M_j(f) \leqslant M_i(f)$$

for $j \leqslant i$, and, consequently, (3.3) implies that

$$d(f) \leqslant \sum_{j \leqslant i} 2^{-j} M_i(f) + \sum_{j > i} 2^{-j},$$

which gives (3.6).

To complete the proof of proposition 3.1, we still have to show that the sets $V(K, \varepsilon)$ form a fundamental system of neighbourhoods of 0 in the topology defined by the above metric.

1) Any set $V(K, \varepsilon)$ is a neighbourhood of 0: for, given K and ε with $\varepsilon < 1$, let i be such that $K \subset K_i$. Then the relation $d(f) \leqslant 2^{-i}\varepsilon$ implies $f \in V(K, \varepsilon)$ because of the first inequality (3. 6).

2) Any neighbourhood of 0 of the form $d(f) \leqslant \varepsilon$ contains a set of the form $V(K, \varepsilon')$. For, given ε, choose an integer i so that $2^{-i} \leqslant \frac{\varepsilon}{2}$; then $f \in V\left(K_i, \frac{\varepsilon}{2}\right)$ implies $d(f) \leqslant \varepsilon$, because of the second inequality (3. 6). Hence, proposition 3. 1 is proved.

Note. We can apply the known properties of metric spaces, or, more precisely, of metrizable topological spaces, to $\mathcal{C}(D)$ and its subspace $\mathcal{H}(D)$. For example, a necessary and sufficient condition for a subset A of a metrizable space E to be *closed* is that each point of E which is the limit of a sequence of points of A belongs to A. Similarly, a necessary and sufficient condition for a mapping f of E into a metrisable space E′ to be *continuous* at a point $x \in E$ is that, for any sequence of points $x_n \in E$ having x as their limit, the sequence $f(x_n)$ has $f(x)$ as its limit. (The reader can refer to *Cours de mathématiques* I of J. Dixmier, *Topologie*, chapter II, § 3.)

We see from the above note that the space $\mathcal{C}(D)$ is *complete*, since the limit of any sequence of continuous functions, which converges uniformly on compact subsets, is continuous. Moreover theorems 1 and 2 of no. 2 can be restated as follows:

The subspace $\mathcal{H}(D)$ is closed in $\mathcal{C}(D)$; the mapping of $\mathcal{H}(D)$ into $\mathcal{H}(D)$ which associates with each function f its derivative f' is continuous.

2. Series of Meromorphic Functions

1. Convergence of a series of meromorphic functions

Let D be an open set of the complex plane $\mathbf{C}$; we shall consider a sequence of meromorphic functions f_n in D. We must define what we mean by the convergence of the series $\sum_n f_n$.

Definition. We say that the series $\sum f_n$ *converges uniformly on the subset* $A \subset D$ if it is possible to remove a finite number of terms from the series in such a way that the remaining functions f_n have no pole in A and form a uniformly convergent series in A.

Similarly, the series $\sum_n f_n$ is said to *converge normally on* A if it is possible to remove a finite number of terms in such a way that the remaining terms f_n have no pole in A and form a normally convergent series in A.

It is clear that any normally convergent series in A is uniformly convergent on A. In what follows, we consider series of meromorphic functions in D which *converge uniformly* (*resp. normally*) *on compact subsets* K *of* D. The sum of such a series in a relatively compact open subset U of D is defined to be the meromorphic function

$$\sum_{n \leqslant n_0} f_n + \Big(\sum_{n > n_0} f_n\Big), \tag{1.1}$$

where n_0 is chosen so that the series $\sum_{n>n_0} f_n$ is uniformly convergent on the closure $\overline{\mathrm{U}}$. The first term in (1.1) is a meromorphic function in U, being the sum of a finite number of meromorphic functions; the second term is a holomorphic function in U, since it is the sum of a uniformly convergent series of holomorphic functions in U. It is easy to see that the meromorphic function defined by (1.1) in U does not depend on the choice of the integer n_0.

THEOREM. *Let* $\sum_n f_n$ *be a series of meromorphic functions* f_n *in* D. *If this series is uniformly* (*resp. normally*) *convergent on compact subsets of* D, *then the sum* f *of this series is a meromorphic function in* D; *the series* $\sum_n f'_n$ *of derivatives converges uniformly* (*resp. normally*) *on compact subsets of* D, *and its sum is the derivative* f' *of the sum* f *of the given series.*

Proof. We have already seen that the sum f is meromorphic in any relatively compact open subset U ⊂ D, so it is meromorphic in D.
Let U be a given relatively compact open subset and let n_0 be chosen as in (1.1); then, in U,

$$f' = \sum_{n \leqslant n_0} f'_n + \Big(\sum_{n > n_0} f_n\Big)'. \tag{1.2}$$

However, the series $\sum_{n>n_0} f_n$ of holomorphic functions can be differentiated term by term because it is uniformly convergent on compact subsets of U; by theorem 2 of § 1, no. 2, the series $\sum_{n>n_0} f'_n$ of derivatives converges uniformly to $\Big(\sum_{n>n_0} f_n\Big)'$ on any compact subset of U. This proves that the series $\sum_n f'_n$ of meromorphic functions converges uniformly to f' on compact subsets of U. Since this is true for any relatively compact open subset U, it follows that $\sum_n f'_n$ converges uniformly to f' on any compact subset of D. If the series $\sum_n f_n$ converges *normally* on compact subsets of D, the fact that the series $\sum_n f'_n$ converges normally on any compact subset of D follows from the lemma of § 1, no. 2.

Note. It is obvious that the set $P(f)$ of poles of f is contained in the union of the sets $P(f_n)$, where $P(f_n)$ denotes the set of poles of f_n. Moreover, relation (1. 1) shows that, if the sets $P(f_n)$ are disjoint from one another, the set $P(f)$ is *equal* to the union of the sets $P(f_n)$; more precisely, if z_0 is a pole of order k of f_n, then it is a pole of order k of f.

2. First example of a series of meromorphic functions

Consider the series

$$\sum_{-\infty<n<+\infty} \frac{1}{(z-n)^2}, \tag{2. 1}$$

the summation extending over all integers n. We shall show that this series converges normally on any compact subset of the plane **C**. Any such compact subset is contained in a strip of the form $x_0 \leqslant x \leqslant x_1$ (we have put $z = x + iy$). It is sufficient then to show that the series (2. 1) converges normally on any strip of the above form. Such a strip only contains a finite number of integers n; in the series $\sum_{n<x_0} \frac{1}{(z-n)^2}$, each term is bounded above by $\frac{1}{(x_0-n)^2}$, and, consequently, this partial series is normally convergent in the strip. Similarly, the partial series $\sum_{n>x_1} \frac{1}{(z-n)^2}$ converges normally on the strip. After removing a suitable finite number of terms of the series (2. 1), we are left with a series of holomorphic functions which are normally convergent on the strip.
This completes the proof.
Let $f(z)$ be the sum of the series (2. 1); it is a meromorphic function in the whole **C**. The function f has period 1:

$$f(z+1) = f(z);$$

for,

$$\sum_n \frac{1}{(z+1-n)^2} = \sum_{n'} \frac{1}{(z-n')^2}, \quad \text{by putting} \quad n-1=n'.$$

The poles of f are the integer points $z = n$ and they are all *double* poles.

The residue at such a pole is zero because, in some neighbourhood of the point $z = n$,

$$f(z) = \frac{1}{(z-n)^2} + g(z) \qquad \text{with } g \text{ holomorphic.}$$

PROPOSITION 2. 1. *The sum $f(z)$ of the series (2. 1) is equal to* $\left(\frac{\pi}{\sin \pi z}\right)^2$.

Proof. The function $f(z)$ tends to 0 as $|y|$ tends to $+\infty$, uniformly with respect to x : in other words, for any $\varepsilon > 0$, there exists an a such that $|y| \geqslant a$ implies $|f(z)| < \varepsilon$. For, suppose first that z remains in a strip $x_0 \leqslant x \leqslant x_1$ and that its imaginary part y satisfies $|y| \geqslant a$ for some $a > 0$; in this domain, the series (2. 1) is a normally convergent series of holomorphic functions; when $|y| \to +\infty$, each term of the series tends to 0 uniformly with respect to x in the strip. Thus the sum of this series (which is normally convergent) tends to 0 as $|y| \to +\infty$ uniformly with respect to x in the strip. However, $f(z)$ has period 1, so, by applying the above property to a strip of width at least 1, we see that $f(z)$ tends to 0 as $|y| \to +\infty$, uniformly with respect to x.

The function $g(z) = \left(\frac{\pi}{\sin \pi z}\right)^2$ has the same following properties as the function $f(z)$:

1° it is meromorphic in **C** and has period 1;

2° its poles are the integers $z = n$ which are double poles with principal part $\frac{1}{(z-n)^2}$;

3° $g(z)$ tends to 0 uniformly with respect to x as $|y| \to +\infty$.

Property 1° is obvious; to prove 2° it is sufficient, because of the periodicity, to show that the origin 0 is a double pole with principal part $\frac{1}{z^2}$ but,

$$(2.\ 2) \quad \left(\frac{\pi}{\sin \pi z}\right)^2 = \left(\frac{\pi}{\pi z - \frac{1}{6}\pi^3 z^3 + \cdots}\right)^2 = \frac{1}{z^2}\left(1 - \frac{1}{6}\pi^2 z^2 + \cdots\right)^{-2}$$

$$= \frac{1}{z^2} + \frac{\pi^2}{3} + z^2(\ldots).$$

Finally, property 3° follows from the relation

$$|\sin \pi z|^2 = \sin^2 \pi x + \sinh^2 \pi y,$$

which shows that $|\sin \pi z|$ tends to infinity (uniformly with respect to x) as $|y|$ tends to infinity.

We can now prove proposition 2. 1 : the function $f(z) - g(z)$ is holomorphic in **C** because f and g have the same poles with the same principal parts. We shall show that $f - g$ is bounded : consider a strip $x_0 \leqslant x \leqslant x_1$; it is bounded for $|y| \leqslant a$ (because a continuous function is bounded on a compact set) and it is bounded for $|y| \geqslant a$ because it tends to 0 as $|y|$ tends to infinity; since it is bounded in each strip, the function $f-g$ is bounded in the whole plane because of its periodicity. Liouville's theorem (chapter III, § 1, no. 2) gives that the function $f-g$ must therefore be constant; since $f-g$ tends to 0 as $|y|$ tends to infinity, this constant is zero. Hence, proposition 2. 1 is proved.

Application. We have

$$\left(\frac{\pi}{\sin \pi z}\right)^2 - \frac{1}{z^2} = \sum_{n\neq 0} \frac{1}{(z-n)^2}, \tag{2. 3}$$

and the right hand side is a holomorphic function $h(z)$ in some neighbourhood of $z = 0$. Moreover, $h(0) = \sum_{n\neq 0} \frac{1}{n^2}$. Hence,

$$\lim_{z\to 0} \left[\left(\frac{\pi}{\sin \pi z}\right)^2 - \frac{1}{z^2}\right] = 2 \sum_{n\geqslant 1} \frac{1}{n^2}. \tag{2. 4}$$

However, the left hand side of (2. 4) is easily evaluated by means of the limited expansion (2. 2); its value is $\frac{\pi^2}{3}$, so we obtain the relation

$$\sum_{n\geqslant 1} \frac{1}{n^2} = \frac{\pi^2}{6} \tag{2. 5}$$

due to Euler.

3. Second example of a series of meromorphic functions

Consider the series

$$\frac{1}{z} + \sum_{n\neq 0} \left(\frac{1}{z-n} + \frac{1}{n}\right). \tag{3. 1}$$

Its general term is equal to $\frac{z}{n(z-n)}$; the reader is left to prove for himself that this series is normally convergent on compact subsets of the plane $\mathbf{C}$. Its sum $F(z)$ is, then, a meromorphic function in $\mathbf{C}$, and its poles are the integers $z = n$; they are simple poles with residues equal to 1. By the theorem of no. 1, the derivative $F'(z)$ is the sum of the series of derivatives, that is to say

$$F'(z) = -\frac{1}{z^2} - \sum_{n\neq 0} \frac{1}{(z-n)^2} = -\left(\frac{\pi}{\sin \pi z}\right)^2 = \frac{d}{dz}\left(\frac{\pi}{\tan \pi z}\right).$$

It follows that $F(z) - \frac{\pi}{\tan \pi z}$ is a constant. However, we see from (3. 1)

that $F(-z) = -F(z)$; thus, the function $F(z) - \frac{\pi}{\tan \pi z}$ is an odd function of z, and, as it is a constant, the constant must be zero.
The series (3. 1) can be rearranged by putting together the two terms corresponding to the integers n and $-n$: since

$$\left(\frac{1}{z-n} + \frac{1}{n}\right) + \left(\frac{1}{z+n} - \frac{1}{n}\right) = \frac{2z}{z^2 - n^2},$$

we finally obtain the relation

$$\frac{1}{z} + \sum_{n \geq 1} \frac{2z}{z^2 - n^2} = \frac{\pi}{\tan \pi z}. \tag{3. 2}$$

4. Another example

By the same method as no. 2, it can be shown that

$$\sum_{-\infty < n < +\infty} \frac{(-1)^n}{(z-n)^2} = \frac{\pi^2}{(\sin \pi z)(\tan \pi z)}; \tag{4. 1}$$

from this, it can be shown by the method of no. 3 that

$$\frac{1}{z} + \sum_{n \geq 1} (-1)^n \frac{2z}{z^2 - n^2} = \frac{\pi}{\sin \pi z}. \tag{4. 2}$$

5. Weierstrass $\wp$ function

Consider, as in chapter III, § 5, no. 5, a discrete subgroup Ω of $\mathbf{C}$ with a base consisting of two vectors e_1 and e_2 whose ratio is not real. We note immediately that the base (e_1, e_2) is not completely determined by the choice of Ω. The vectors of another base (e'_1, e'_2) are expressible as linear combinations of the vectors of the first base with integral coefficients, and vice-versa; it follows that the determinant of the matrix of coefficients is an integer which has an inverse in the ring of integers, so it is equal to ± 1. Conversely, if e'_1 and e'_2 are linear combinations of e_1 and e_2 with integral coefficients and if the determinant of the matrix of coefficients is equal to ± 1, then the Cramer formulae show that, conversely, e_1 and e_2 are linear combinations of e'_1 and e'_2 with integer coefficients, and, consequently, (e'_1, e'_2) is a base of Ω.

Proposition 5. 1. *Given a discrete subgroup Ω as above, the series*

$$\wp(z) = \frac{1}{z^2} + \sum_{\omega \in \Omega - \{0\}} \left(\frac{1}{(z-\omega)^2} - \frac{1}{\omega^2}\right) \tag{5. 1}$$

is normally convergent on compact subsets of the plane $\mathbf{C}$.

To prove this we shall need the following lemma :

Lemma. *The series* $\sum_{\omega\in\Omega,\ \omega\neq 0} \frac{1}{|\omega|^3}$ *is convergent.*

Proof of the lemma. For each integer $n \geqslant 1$, consider the parallelogramm P_n formed by the points $z = t_1 e_1 + t_2 e_2$ where the real numbers t_1 and t_2 are such that $\sup(|t_1|, |t_2|) = n$ (cf. figure 10). There are exactly $8n$

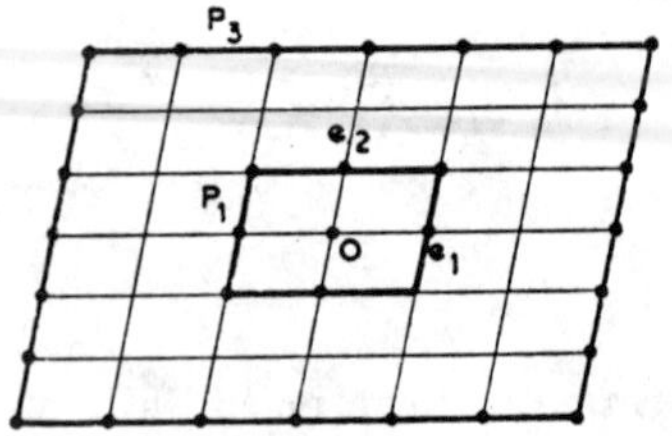

Fig. 10

points of Ω on P_n, and the distance between any of them and o is $\geqslant kn$, where k is a fixed number > 0 (it is the smallest distance from o to the points of P_1). The sum of the $\frac{1}{|\omega|^3}$ taken over the points of P_n is therefore bounded above by $\frac{8n}{k^3 n^3}$, so

$$\sum_{\omega\neq 0} \frac{1}{|\omega|^3} \leqslant \sum_{n\geqslant 1} \frac{8}{k^3 n^2},$$

and, since the series $\sum \frac{1}{n^2}$ is convergent, the lemma is proved.

We can now show that the series (5. 1) converges normally on any compact disc $|z| \leqslant r$. We have $|\omega| \geqslant 2r$ for all but a finite number of the ω; thus, for all but a finite number of the terms of the series (5. 1),

$$\left|\frac{1}{(z-\omega)^2} - \frac{1}{\omega^2}\right| = \left|\frac{2\omega z - z^2}{\omega^2(\omega - z)^2}\right| = \frac{\left|z\left(2 - \frac{z}{\omega}\right)\right|}{|\omega^3|\left|1 - \frac{z}{\omega}\right|^2} \leqslant \frac{r \cdot \frac{5}{2}}{|\omega|^3 \cdot \frac{1}{4}} = \frac{10\,r}{|\omega|^3} \text{ as } |z| \leqslant r.$$

It follows from the lemma that the series (5. 1) converges normally in the disc $|z| \leqslant r$.

Definition. The Weierstrass function $\wp(z)$ is defined to be the meromorphic function which is the sum of the series (5. 1). (This function depends, of course, on the discrete subgroup Ω chosen).

The poles of $\wp$ are exactly the points of Ω; they are double poles whose residue is zero : for, in some neighbourhood of $z = \omega$,

$$\wp(z) = \frac{1}{(z-\omega)^2} + g(z), \text{ where } g \text{ is holomorphic.}$$

The function $\wp$ is an *even* function of z because

$$\wp(-z) = \frac{1}{z^2} + \sum_{\omega \neq 0} \left(\frac{1}{(z+\omega)^2} - \frac{1}{\omega^2} \right),$$

and it is sufficient to put $-\omega$ for ω on the right hand side to recover the series (5. 1). By the theorem of no. 1, an expansion of the derived function $\wp'$ as a series (which is normally convergent on compact subsets) is

$$\wp'(z) = -2 \sum_{\omega \in \Omega} \frac{1}{(z-\omega)^3}. \tag{5. 2}$$

This relation demonstrates the *periodicity* of the function $\wp'$,

$$\wp'(z+\omega) = \wp'(z) \qquad \text{for all } \omega \in \Omega,$$

and the fact that $\wp'(-z) = -\wp'(z)$.

We shall now show that the function $\wp$ itself has any $\omega \in \Omega$ as period. To do this, it is sufficient to prove that $\wp(z+e_i) = \wp(z)$ when i takes the values 1 and 2. However,

$$\wp(z+e_i) - \wp(z) = \text{constant} \tag{5. 3}$$

because the derivative $\wp'(z+e_i) - \wp'(z) = 0$. We can give z the value $-\frac{e_i}{2}$ in relation (5. 3) because $\frac{e_i}{2}$ and $-\frac{e_i}{2}$ are not poles of $\wp$; we find then that the right hand side of (5. 3) is equal to $\wp\left(\frac{e_i}{2}\right) - \wp\left(-\frac{e_i}{2}\right)$, which is zero since $\wp$ is an even function.

To sum up, the Weierstrass $\wp$-function is a meromorphic function with the points of Ω as periods and with poles at the points of Ω and no others, each pole having order 2 and principal part $\frac{1}{(z-\omega)^2}$.

The Laurent expansion of $\wp(z)$. In some neighbourhood of the origin, $\wp$ has a Laurent expansion which is *a priori*, of the form

$$\wp(z) = \frac{1}{z^2} + a_2 z^2 + a_4 z^4 + \cdots, \tag{5. 4}$$

because the function $\wp$ is even and because, by (5. 1), the holomorphic function defined by

$$g(z) = \wp(z) - \frac{1}{z^2} = \sum_{\omega \neq 0} \left(\frac{1}{(z-\omega)^2} - \frac{1}{\omega^2} \right)$$

in some neighbourhood of the origin, is zero for $z = 0$. The coefficients a_2 and a_4 can easily be expressed in terms of the discrete subgroup Ω; differentiating the series $g(z)$ term by term gives

$$(5.5) \qquad a_2 = 3 \sum_{\omega \neq 0} \frac{1}{\omega^4}, \qquad a_4 = 5 \sum_{\omega \neq 0} \frac{1}{\omega^6}.$$

Now, differentiate relation (5. 4) term by term and square both sides; this gives

$$(5.6) \qquad (\wp'(z))^2 = \frac{4}{z^6} - \frac{8a_2}{z^2} - 16 a_4 + \cdots;$$

cubing both sides of (5. 4) gives

$$(5.7) \qquad (\wp(z))^3 = \frac{1}{z^6} + \frac{3a_2}{z^2} + 3a_4 + \cdots,$$

whence,

$$\wp'^2 - 4\wp^3 = -20 \frac{a_2}{z^2} - 28a_4 + z^2(\cdots).$$

It follows that the function

$$(5.8) \qquad \wp'^2 - 4\wp^3 + 20a_2\wp + 28a_4$$

is holomorphic in some neighbourhood of the origin and is zero at the origin. However, this function has Ω as its group of periods, so it is holomorphic in some neighbourhood of each point of Ω and zero at each point of Ω. Since the function has no poles outside Ω, it follows that it is holomorphic in the whole plane; since it is bounded on any compact subset, its periodicity implies that it is bounded on $\mathbf{C}$; and, since it is zero at the origin, Liouville's theorem shows that it is identically zero. Finally, we have proved

$$(5.9) \qquad \wp'^2 - 4\wp^3 + 20a_2\wp + 28a_4 = 0.$$

This relation has an important interpretation : let us consider the algebraic curve

$$(5.10) \qquad y^2 = 4x^3 - 20a_2x - 28a_4;$$

the formulae $x = \wp(z)$, $y = \wp'(z)$ give a parametric representation of this curve. We shall show that any point $(x, y) \in \mathbf{C} \times \mathbf{C}$ which satisfies (5. 10) is the image of a point $z \in \mathbf{C}$, which is determined up to the addition of an element of Ω.

First we seek the $z \in \mathbf{C}$ such that $2z \in \Omega$ and $z \notin \Omega$. At such a point, $\wp$ and $\wp'$ are holomorphic; we have $\wp'(z) = \wp'(-z)$ because of the periodicity of $\wp'$ and $\wp'(z) = -\wp'(-z)$ because of $\wp'$ is an odd function; hence, $\wp'$ is zero at such a point. We know three such points, namely,

$$(5.11) \qquad z = e_1/2, \quad z = e_2/2, \quad z = (e_1 + e_2)/2,$$

and we immediately see that any z such that $2z \in \Omega$ and $z \notin \Omega$ is congruent (mod. Ω) to one of the three points (5. 11); the classes (mod. Ω) of the three points (5. 11) are distinct. Since $\wp'$ has a unique triple pole in each parallelogram of periods, proposition 5. 1 of chapter III, § 5, shows that $\wp'$ has, at the most, three distinct zeros in each parallelogram of periods. These are therefore the three points (5. 11) or their conjugates (mod. Ω). By the same proposition, the function $\wp$ does not take a given value more than twice in each parallelogram of periods. Since $\wp(z_0) = \wp(-z_0)$, the function takes each value of the form $\wp(z_0)$ exactly twice if $2z_0 \notin \Omega$; on the other hand, if $2z_0 \in \Omega$ and $z_0 \notin \Omega$, then $\wp'(z_0) = 0$ as we have just seen, and the equation $\wp(z) = \wp(z_0)$ has z_0 as a double root, so $\wp$ only takes the value $\wp(z_0)$ once in each parallelogram of periods. From these results, it follows that each of the values

$$\wp(e_1/2), \quad \wp(e_2/2), \quad \wp((e_1 + e_2)/2)$$

is taken precisely once in each parallelogram of periods and that these three values are distinct. By (5. 9), they are the three roots of the equation

$$4x^3 - 20a_2x - 28a_4 = 0, \tag{5. 12}$$

and, consequently, this equation has three distinct roots. To sum up, we have proved:

PROPOSITION 5. 2. *Given the discrete group Ω, equation* (5. 12), *whose coefficients a_2 and a_4 are defined by* (5. 5), *has three distinct roots. Moreover, for each point*

$$(x, y) \in \mathbf{C} \times \mathbf{C}$$

of the algebraic curve (5. 10), *there is a unique (modulo Ω) $z \in \mathbf{C}$ such that*

$$\wp(z) = x, \qquad \wp'(z) = y.$$

We shall see (cf. chapter VI, § 5, no. 3) that, conversely, given an arbitrary equation of the form (5. 10) whose right hand side has three distinct roots, there exists a discrete group Ω such that a_2 and a_4 satisfy (5. 5); if $\wp$ is the Weierstrass function associated with this group Ω, the formulae $x = \wp(z)$, $y = \wp'(z)$ give a parametric representation of the algebraic curve (5. 10).

3. Infinite Products of Holomorphic Functions

1. DEFINITIONS.

Definition. Let $(f_n(z))$ be a sequence of continuous functions defined in an open set D of the complex plane. We say that the infinite product $\prod_n f_n(z)$ *converges normally* on a subset $K \subset D$ if the following two conditions are satisfied :

1° as n tends to $+\infty$, $f_n(z)$ tends uniformly to 1 on K; this implies in particular that, for sufficiently large n, $f_n - 1$ has modulus < 1 on K and, consequently, $\log f_n$ is a function defined in K (we take the principal branch of the logarithm);

2° the series whose general term is $\log f_n$ (which is defined for sufficiently large n) is normally convergent in K.

We can give a single condition which is equivalent to the two conditions 1° and 2° above. Let $f_n = 1 + u_n$; then, condition 1° expresses that the sequence u_n converges uniformly to 0 in K; when u_n is small, $\log f_n$ and u_n are equivalent to the first order, and, consequently, condition 2° expresses that the series $\sum_n u_n$ converges normally in K. To sum up, *a necessary* and *sufficient condition for the infinite product* $\prod_n f_n$ *to converge normally in* K *is that the series* $\sum_n u_n$ *converges normally in* K.

Definition. The infinite product $\prod_n f_n$ is said to *converge normally on compact subsets* of D if this product converges normally an every compact subset K of D.

A necessary and sufficient condition for this is that, if we put $f_n = 1 + u_n$, the series $\sum_n u_n$ converges normally on compact subsets of D. If this is so, then, as n_0 increases indefinitely, the products $\prod_{n \leq n_0} f_n$ converge uniformly on compacts subsets of D to a limit $f(z)$, which is evidently a continuous function of z. To see this convergence, it is sufficient to take logarithms of the factors f_n for sufficiently large n.

2. Properties of normally convergent products of holomorphic functions

THEOREM 1. *If the functions f_n are holomorphic in* D *and if the infinite product* $\prod_n f_n$ *converges normally on compact subsets of* D, *then* $f = \prod_n f_n$ *is holomorphic in* D. *Moreover,*

$$f = f_1 f_2 \cdots f_p \left(\prod_{n>p} f_n\right). \tag{2. 1}$$

The set of zeros of f is the union of the sets of zeros of the functions f_n, the order of multiplicity of a zero of f being equal to the sum of the orders of multiplicity which it has for each of the functions f_n.

Proof. f is holomorphic because f is the limit (approached uniformly on compact subsets) of the finite products, which are holomorphic. The associativity formula (2. 1) is obvious on any relatively compact open set U The function f_n has no zeros in U for sufficiently large n, since $u_n = f_n - 1$ converges uniformly to 0 in U; the last statement is then obvious.

Theorem 2. *With the hypotheses of theorem 1, the series* $\sum_n f'_n/f_n$ *of meromorphic functions converges normally on compact subsets of* D *(in the sense of* no. 1 *of* § 2) *and its sum is merely the logarithmic derivative* f'/f.

Proof. Let U be a relatively compact open set of D. The function

$$g_p = \exp\Big(\sum_{n>p} \log f_n\Big) \tag{2.2}$$

is defined and holomorphic in U for sufficiently large p. By (2. 1), we have, in U,

$$\frac{f'}{f} = \sum_{n\leqslant p} \frac{f'_n}{f_n} + \frac{g'_p}{g_p}. \tag{2.3}$$

However,

$$\frac{g'_p}{g_p} = \sum_{n>p} \frac{f'_n}{f_n}, \tag{2.4}$$

where the series on the right hand side converges uniformly on compact subsets of D; for, the series $\sum_{n>p} \log f_n$ of logarithms converges (uniformly on compact subsets) to $\log g_p$, so the series of derivatives of these logarithms converges (uniformly on compact subsets) to the derivative g'_p/g_p (cf. § 1, no. 2, theorem 2). By comparing (2. 3) and (2. 4), we see that, in U,

$$\frac{f'}{f} = \sum_n \frac{f'_n}{f_n},$$

the convergence being normal on compact subsets of U. This holds for all U, whence the theorem.

3. Example : expansion of sin πz as an infinite product

Consider the infinite product

$$f(z) = z\prod_{n\geqslant 1}\left(1 - \frac{z^2}{n^2}\right). \tag{3.1}$$

This product converges normally on compact subsets of the plane $\mathbf{C}$ because the series $\sum_n \frac{z^2}{n^2}$ converges normally on compact subsets, which follows from the convergence of the numerical series $\sum_n \frac{1}{n^2}$. Thus, $f(z)$ is a holomorphic function in the whole plane and its zeros are all the integral values of z. They are simple zeros.

By theorem 2, we can differentiate logarithmically term by term; we obtain a series of meromorphic functions, which converges normally on compact subsets of the plane,

$$(3.2)\qquad \frac{f'(z)}{f(z)} = \frac{1}{z} + \sum_{n\geqslant 1} \frac{2z}{z^2 - n^2}.$$

We have seen (§ 2, no. 3) that the sum of this series is

$$\frac{\pi}{\tan \pi z} = \frac{g'(z)}{g(z)},$$

where $g(z) = \sin \pi z$. Thus, $f'/f = g'/g$, so

$$\frac{f(z)}{z} = c\frac{\sin \pi z}{z}.$$

The constant c remains to be determined. By (3. 1), $f(z)/z$ tends to 1, as z tends to 0, and, since $\frac{\sin \pi z}{z}$ has π as its limit, we see that $c = \frac{1}{\pi}$. Hence, we have proved the formula

$$(3.3)\qquad \frac{\sin \pi z}{\pi z} = \prod_{n\geqslant 1}\left(1 - \frac{z^2}{n^2}\right).$$

4. The Γ-function

Consider the holomorphic function g_n defined for each integer $n \geqslant 1$ by

$$(4.1)\qquad g_n(z) = z(1+z)\left(1+\frac{z}{2}\right)\cdots\left(1+\frac{z}{n}\right)n^{-z}$$
$$= \frac{z(z+1)(z+2)\ldots(z+n)}{n!}n^{-z}.$$

We have, for $n \geqslant 2$,

$$(4.2)\qquad \frac{g_n(z)}{g_{n-1}(z)} = \left(1+\frac{z}{n}\right)\left(1-\frac{1}{n}\right)^z = f_n(z).$$

If $|z| \leqslant r$ and $1 \leqslant r < n$, we can consider the principal branch of $\log f_n(z)$ and so

$$(4.3)\qquad |\log f_n(z)| \leqslant 2\left(\frac{r^2}{2n^2} + \frac{r^3}{3n^3} + \cdots\right) \leqslant 2\frac{r^2}{n^2}$$

for sufficiently small $\frac{r}{n}$. Hence, the series $\sum_n \log f_n(z)$ converges normally on compact subsets of the plane, and, consequently, the infinite product $g_1 \cdot \prod_{n\geqslant 2} \frac{g_n}{g_{n-1}}$ converges normally on compact subsets of the plane. Its value is a holomorphic function $g(z)$, which is the uniform limit on compact subsets of the functions

$$g_n = g_1 f_2 \ldots f_n.$$

The function g has the numbers $0, -1, -2, \ldots, -n, \ldots$ as its zeros, and they are all simple zeros. If z is not an integer, we can form the quotient

$$(4.4)\qquad \frac{g(z)}{g(z+1)} = \lim_{n\to\infty} \frac{g_n(z)}{g_n(z+1)} = \lim_{n\to\infty} \frac{nz}{n+z+1} = z.$$

Thus, the meromorphic function $\frac{g(z)}{g(z+1)}$ is, in fact, holomorphic and identical with z. Moreover,

$$(4.5)\qquad g(1) = \lim_{n\to\infty} g_n(1) = \lim_{n\to\infty} \frac{n+1}{n} = 1.$$

Definition. The meromorphic function $1/g(z)$ is denoted by $\Gamma(z)$. All the integers $n \leqslant 0$ are simple poles of $\Gamma(z)$, and the function satisfies

$$(4.6)\qquad \Gamma(z+1) = z\Gamma(z), \qquad \Gamma(1) = 1,$$

which follows obviously from (4.4) and (4.5). We deduce from (4.6) by induction on the integer $n \geqslant 0$ that

$$(4.7)\qquad \Gamma(n+1) = n!$$

We now propose to calculate the product $\Gamma(z).\Gamma(1-z)$. We have

$$(4.8)\qquad g(z).g(1-z) = \lim_{n\to\infty} \frac{n+1-z}{n}.\ z.\prod_{k=1}^{n}\left(1-\frac{z^2}{k^2}\right),$$

which, by no.3, is equal to $\frac{\sin \pi z}{\pi}$. By inverting, we obtain

$$(4.9)\qquad \Gamma(z).\Gamma(1-z) = \frac{\pi}{\sin \pi z},$$

and, in the particular case when $z = \frac{1}{2}$,

$$\Gamma\left(\frac{1}{2}\right) = \sqrt{\pi}.$$

The Weierstrass infinite product. By using (4.1), we can write

$$(4.10)\qquad g_n(z) = z.\prod_{k=1}^{n}\left(\left(1+\frac{z}{k}\right)e^{-z/k}\right).e^{z\left(1+\frac{1}{2}+\cdots+\frac{1}{n}-\log n\right)}.$$

The exponent $z\left(1+\cdots+\frac{1}{n}-\log n\right)$ tends to Cz as n tends to infinity, where C denotes Euler constant. In the limit, we obtain

$$(4.11)\qquad g(z) = ze^{Cz}\prod_{k=1}^{\infty}\left(\left(1+\frac{z}{k}\right)e^{-z/k}\right),$$

and the reader can verify that the product on the right hand side is normally convergent on compact subsets of the plane. Since $g = 1/\Gamma$, we obtain, by taking logarithmic derivatives of the two side of (4. 11) (cf. theorem 2),

$$\frac{\Gamma'(z)}{\Gamma(z)} = -\frac{1}{z} - C + \sum_{n \geqslant 1} \left(\frac{1}{n} - \frac{1}{z+n}\right), \tag{4. 12}$$

and, in particular,

$$-C = \lim_{z \to 0} \left(\frac{\Gamma'(z)}{\Gamma(z)} + \frac{1}{z}\right). \tag{4. 13}$$

Finally, we can differentiate relation (4. 12) term by term (cf. § 2, no. 1) to obtain

$$\frac{d}{dz}\left(\frac{\Gamma'(z)}{\Gamma(z)}\right) = \sum_{n \geqslant 0} \frac{1}{(z+n)^2}. \tag{4. 14}$$

Note the similarity between the series on the right hand side and the series whose sum is $\left(\frac{\pi}{\sin \pi z}\right)^2$ (§ 2, no. 2). When z is *real and positive*, the right hand side of (4. 14) is obviously positive, so $\log \Gamma(z)$ *is a convex function of* z for real $z > 0$.

4. Compact Subsets of $\mathcal{H}(D)$

The characterization of compact subsets of $\mathcal{H}(D)$ which is given here is what used to be called the theory of 'normal families' of holomorphic functions.

1. Bounded subsets of $\mathcal{H}(D)$

We shall define what is meant by bounded subsets of the vector space $\mathcal{H}(D)$. The definition is merely a particular case of a definition which applies to any topological vector space. In particular the same definition applies to bounded subsets of $\mathcal{C}(D)$.

Definition. A subset $A \subset \mathcal{H}(D)$ is *bounded* if, for any neighbourhood $V(K, \varepsilon)$ of 0, there is a finite positive number λ such that $A \subset \lambda V(K, \varepsilon)$, where $\lambda V(K, \varepsilon)$ denotes the homothety of $V(K, \varepsilon)$ with respect to the origin 0 by the factor λ. The relation $A \subset \lambda V(K, \varepsilon)$ expresses that $|f(z)| \leqslant \lambda\varepsilon$

for $z \in K$ and for any function $f \in A$. Hence, *a necessary and sufficient condition for a set* A *of holomorphic functions in* D *to be bounded is that, for any compact subset* $K \subset D$, *there exists a finite number* M(K) *such that*

$$(1.1) \qquad |f(z)| \leqslant M(K) \quad \textit{for all} \quad z \in K \textit{ and all } f \in A.$$

In other words, A is a bounded subset if the functions $f \in A$ are *uniformly bounded on compact subsets of* D (the upper bound M(K) depending clearly on the compact subset K).

If A is a bounded subset of $\mathcal{H}(D)$, its closure $\bar{A}$ is bounded (we mean here the closure in the topology of uniform convergence on compact subsets of D). This is obvious because, if (1.1) holds for every $f \in A$, it holds also for every function belonging to the closure of A.

PROPOSITION 1.1. *The mapping* $f \to f'$ *of* $\mathcal{H}(D)$ *into itself takes any bounded subset into a bounded subset.*

This follows directly from the lemma which we used to prove theorem 2 of § 1, no. 2.

2. STATEMENT OF THE FUNDAMENTAL THEOREM

We propose to characterize the *compact* subsets of the space $\mathcal{H}(D)$ of holomorphic functions in an open set D of the complex plane.

PROPOSITION 2.1. *If* $A \subset \mathcal{H}(D)$ *is compact, then* A *is closed and bounded.*

Proof. The space $\mathcal{H}(D)$ is Hausdorff since it is metrizable (cf. § 1, no. 3). Thus, any compact subset of $\mathcal{H}(D)$ is closed by a classical result in general topology. It remains to be proved that, if A is compact, A is bounded, To prove this, let K be a compact subset of D and consider the mapping

$$f \to \sup_{z \in K} |f(z)|$$

of the space $\mathcal{H}(D)$ into $\mathbf{R}$; it is clear that this is a continuous mapping, so the set of values which it takes on the compact subset consisting of the $f \in A$ is bounded. This expresses that the $f \in A$ are uniformly bounded on the compact subset K. It holds for any compact subset K of D, and, consequently, the set A is indeed a bounded subset of the vector space $\mathcal{H}(D)$.

Note. Proposition 2.1 is stated for the space $\mathcal{H}(D)$ but it is also true for the space $\mathcal{C}(D)$ of continuous functions in D. In constrast, the converse of proposition 2.1, which we shall now state, is only true for subsets of the space $\mathcal{H}(D)$ of *holomorphic* functions in D.

FUNDAMENTAL THEOREM. *Any subset of $\mathcal{H}$(D) which is bounded and closed is compact.*

COROLLARY. *A necessary and sufficient condition for a subset* A *of* $\mathcal{H}$(D) *to be compact is that it is bounded and closed.*

The proof of this theorem will take up numbers 3, 4 and 5. An equivalent statement of the fundamental theorem is the following :

Any bounded subset of $\mathcal{H}$(D) *is relatively compact.* The converse is also true.

3. METHOD OF PROOF OF THE FUNDAMENTAL THEOREM.

Let A be a bounded, closed subset of $\mathcal{H}$(D). The topological space A is metrizable since it is a subspace of a metrizable space $\mathcal{H}$(D). To prove that A is compact it is sufficient to show that *any infinite sequence of elements of* A *has an infinite sub-sequence which converges to an element of* A. For, we have the following topological lemma :

LEMMA 1. *Let* A *be a metric space such that any infinite sequence of points of* A *contains an infinite subsequence which converges to a point of* A; *then* A *is compact.*

Proof of lemma 1. Let (U_i) be a covering of A by open sets U_i. We must show that this covering contains a finite covering.
First we show :

a) There exists an $\varepsilon > 0$ such that any ball $B(x, \varepsilon)$ is contained in at least one of the U_i (we use $B(x, \varepsilon)$ to denote the closed ball of centre $x \in A$ and radius ε).

To prove *a*) we use *reductio ad absurdum* : we assume that there is a sequence of points $x_n \in A$ and a decreasing sequence of numbers ε_n tending to zero such that, for each n, the ball $B(x_n, \varepsilon_n)$ is not contained in any of the U_i. By hypothesis, the sequence (x_n) contains an infinite subsequence which converges to a point $a \in A$. We can therefore suppose that the sequence (x_n) converges to a. Let U_i be an open set containing a; then U_i contains a ball $B(a, r)$. For sufficiently large n, $x_n \in B(a, r/2)$ and $\varepsilon_n \leqslant r/2$. It follows that $B(x_n, \varepsilon_n)$ is contained in U_i for sufficiently large n, which is a contradiction. This proves *a*). We now show :

b) For any $\varepsilon > 0$, A can be covered by a *finite* number of balls $B(x_n, \varepsilon)$.
It is clear that *a*) and *b*) together imply that there exists a finite number of the open sets U_i which cover A.

We prove *b*) also by seeking a contradiction : if *b*) is false, there is an infinite sequence of points $x_n \in A$ whose distances apart are all $\geqslant \varepsilon$; however, we can, by hypothesis, extract a convergent subsequence of this sequence, which obviously leads to a contradiction. Thus, we have proved lemma 1.

4. A LEMMA

Because of no. 3, we have reduced the theorem to showing that, if A is a bounded subset of $\mathcal{H}(D)$, then any infinite sequence of functions $f_k \in A$ contains an infinite subsequence which converges uniformly on compact subsets of D. The following lemma is a useful criterion for convergence of sequences of holomorphic functions belonging to a bounded subset :

LEMMA 2. *Let* D *be an open disc centred at* z_0 *and let* A *be a bounded subset of* $\mathcal{H}(D)$. *A necessary and sufficient condition for a sequence of functions* $f_k \in A$ *to be convergent (in the topology of uniform convergence on compact subsets of* D) *is the following condition* :

$C(z_0)$ *for each* $n \geqslant 0$, *the sequence of n-th derivatives* $f_k^{(n)}(z_0)$ *has a limit.* (For $n = 0$ this means that the sequence of values of the functions f_k at the point z_0 has a limit.).

Proof of lemma 2. Condition $C(z_0)$ is necessary because, for each n, the sequence of n-th derivatives $f_k^{(n)}$ converges uniformly on any compact subset of D (§ 1, no. 2, theorem 2). It remains to be proved that condition $C(z_0)$ implies that the sequence (f_k) converges uniformly in any compact disc of centre z_0 and radius r strictly less than the radius of the disc D. Choose an $r_0 > r$ but still strictly less than the radius of D. Since A is bounded, there exists a finite M such that

$$|f_k(z)| \leqslant M \quad \text{for} \quad |z - z_0| \leqslant r_0. \tag{4.1}$$

We consider the Taylor expansions of the holomorphic functions f_k :

$$f_k(z) = \sum_{n \geqslant 0} a_{n,k}(z - z_0)^n. \tag{4.2}$$

By Cauchy's inequalities, we have

$$|a_{n,k}| \leqslant \frac{M}{(r_0)^n}. \tag{4.3}$$

Thus, for $|z - z_0| \leqslant r$ and for all k and h,

$$|f_k(z) - f_h(z)| \leqslant \sum_{0 \leqslant n \leqslant p} |a_{n,k} - a_{n,h}|\, r^n + 2M \sum_{n > p} \left(\frac{r}{r_0}\right)^n. \tag{4.4}$$

Since $r/r_0 < 1$, we can choose p sufficiently large to make

$$2M \sum_{n>p} \left(\frac{r}{r_0}\right)^n$$

less than $\frac{\varepsilon}{2}$, where ε is an arbitrary number > 0 given in advance. By condition $C(z_0)$, when the integers k and h both increase indefinitely, the difference $a_{n,k} - a_{n,k}$ tends to zero for each n because

$$a_{n,k} = \frac{1}{n!} f_k^{(n)}(z_0).$$

We can therefore choose the integer k_0 such that

$$\sum_{0 \leqslant n \leqslant p} |a_{n,k} - a_{n,h}| \, r^n \leqslant \frac{\varepsilon}{2} \quad \text{for} \quad k \geqslant k_0, \quad h \geqslant k_0.$$

Thus, from (4. 4),

$$(4.5) \quad |f_k(z) - f_n(z)| \leqslant \varepsilon \quad \text{for} \quad k \geqslant k_0, h \geqslant h_0, |z - z_0| \leqslant r,$$

which proves that the sequence of functions f_k converges uniformly on the compact disc of centre z_0 and radius r. Hence, lemma 2 is proved.

5. Proof of the fundamental theorem

We are now in a position to prove the fundamental theorem (no. 2).

The given open set D can be covered by a countable sequence of open discs with centres at $z_i \in D$. For each integer $n \geqslant 0$ and for each i, consider the linear mapping

$$(5.1) \qquad \lambda_i^n : \mathcal{H}(D) \to \mathbf{C}$$

which associates the number $f^{(n)}(z_i)$ with each function f. Let us consider a sequence of functions f_k belonging to a bounded subset A; we intend to show that there is an infinite subset N′ of the set of positive integers N such that

$$(5.2) \qquad \lim_{k \in N'} \lambda_i^n(f_k) \quad \text{exists for each pair } (i, n).$$

But, for each i and each n, the numbers $\lambda_i^n(f_k)$ form a *bounded* sequence as the index k describes N, since the f_k describe a bounded set A and the mappings λ_i^n are continuous. Let us arrange the countable set of mappings λ_i^n into a single sequence which we write $\mu_1, \ldots, \mu_m, \ldots$. We shall show the existence of an infinite subset N′ of N such that

$$(5.3) \qquad \lim_{k \in N'} \mu_m(f_k) \quad \text{exists for each integer } m \geqslant 1.$$

To do this, we shall use the diagonal sequence method. Since the sequence of the $\mu_1(f_k)$, for $k \in N$, is bounded, there is an infinite subset $N_1 \subset N$ such that

$$\lim_{k \in N_1} \mu_1(f_k) \quad \text{exists.}$$

The sequence of the $\mu_2(f_k)$, for $k \in N_1$, is bounded, so there exists an infinite subset $N_2 \subset N_1$ such that

$$\lim_{k \in N_2} \mu_2(f_k) \quad \text{exists.}$$

Hence we define, step by step, the infinite subsets

$$N_1 \supset N_2 \supset \ldots \supset N_m \supset \ldots.$$

The set N_{m+1} is then an infinite subset of N_m such that

$$\lim_{k \in N_{m+1}} \mu_{m+1}(f_k) \quad \text{exists.}$$

Consider now the infinite sequence N' of integers defined in the following way : for each integer $m \geqslant 1$, the m-th term of the sequence N' is the m-th term of the sequence N_m. The sequence N' is strictly increasing, and it is clear that every integer of sequence N' after the m-th belongs to N_m. This holds for all m, so the sequence N' satisfies condition (5. 3), which completes the proof.

Hence the fundamental theorem of no. 2 is completely proved.

Note. What we have just proved is, in fact, that, in a special case, an infinite product of compact spaces is compact.

6. Some consequences of the fundamental theorem

The following principle is frequently used : *Let* A *be a bounded set of holomorphic functions in* D; *if a sequence of functions* $f_k \in A$ *has not more than one function in its closure (in the topology of uniform convergence on compact subsets), this sequence is convergent (in the same topology).* This follows from a classical theorem about the topology of compact spaces.

As an application of this principle, consider first the case where the open set D is *connected* and where the sequence of functions f_k converges simply *at each point of a non-empty open set* D' contained in D (this convergence means that the sequence of numbers $f_k(z)$ has a limit for each point $z \in D'$). If this is the case and if the f_k belong to a *bounded* set, the sequence (f_k) converges uniformly on closed subsets of D. For, if f and g are two holomorphic functions in D and both are in the closure of the sequence of the f_k, then, clearly, $f(z) = g(z)$ at any point $z \in D'$ which implies that f and g are identical in D (by the principle of analytic continuation).

Consider now the case of a bounded sequence of holomorphic functions f_k satisfying the conditions $C(z_0)$ of lemma 2, where z_0 is a point of D. Then, if D is connected, the sequence (f_k) converges uniformly in any compact subset of D. For, if f and g are two holomorphic functions in the closure of the sequence (f_k), then $f^{(n)}(z_0) = g^{(n)}(z_0)$ for any integer $n \geqslant 0$, and consequently, f and g are identical because of the principle of analytic continuation.

Another case is that of a bounded sequence of holomorphic functions f_k in D which converges simply at each point of a *non-discrete* subset E of D, where D is still connected. Such a sequence converges uniformly on compact subsets of D, because, if f and g are two holomorphic functions in the closure of the sequence (f_k), the difference $f(z)-g(z)$ is zero at each point of E, and is thus identically zero since the set of zeros of a holomorphic function in D (connected) is a discrete set if the function is not identically zero.

Exercises

1. Let $f(z)$ be a holomorphic function in the disc $|z| < 1$, and suppose that $f(0) = 0$. Show that the series $\sum_{n \geqslant 1} f(z^n)$ converges uniformly in any compact subset of this disc. (Given $0 < r < 1$, use Schwarz' lemma (in the disc $|z| < r$) to bound $|f(z^n)|$ above by a constant multiple of $|z|^n$ for $|z| \leqslant r$).

2. Let D be a connected open set of the plane $\mathbf{C}$ and let $(f_n(z))$ be a sequence of holomorphic functions in D, converging uniformly on compact subsets of D to a function $f(z)$ which is not identically zero. Moreover, let Γ be the oriented boundary of a compact subset K of D such that $f(z) \neq 0$ on Γ. Show that there is a positive integer N such that, for $n \geqslant N, f_n(z) \neq 0$ on Γ, and that f_n and f have the same number of zeros in K. (If M is the lower bound of $|f(z)|$ on Γ and if N is chosen so that $|f_n(z)-f(z)| < M$ for $n \geqslant N$ and $z \in \Gamma$, then Rouché's theorem (exercise 19 of chapter III) can be applied to the functions $f(z)$ and $f_n(z)-f(z)$.)

Deduce that, if a is a zero of $f(z)$, there exists a sequence (a_n) of points of D such that

$$\lim_n a_n = a, \qquad f_n(a_n) = 0.$$

3. Let τ be a complex number such that $\operatorname{Im}(\tau) > 0$, and put $q = e^{\pi i \tau}$. Show that the following two series converge uniformly on compact subsets of the plane $\mathbf{C}$ of the variable u:

$$\sum_{-\infty < n < +\infty} (-1)^n q^{n^2} e^{2\pi n i u},$$

$$-i \sum_{-\infty < n < +\infty} (-1)^n q^{\left(n+\frac{1}{2}\right)^2} e^{(2n+1)\pi i u}.$$

If we denote the holomorphic functions defined (in the whole plane) by these series by $\vartheta_0(u)$, $\vartheta_1(u)$, then the following relations hold :

$$\vartheta_0(u+1) = \vartheta_0(u), \qquad \vartheta_1(u+1) = -\vartheta_1(u),$$
$$\vartheta_0(u+\tau) = -q^{-1}e^{-2\pi iu}\vartheta_0(u), \qquad \vartheta_1(u+\tau) = -q^{-1}e^{-2\pi iu}\vartheta_1(u),$$
$$\vartheta_0\left(u+\frac{\tau}{2}\right) = iq^{-1/4}e^{-\pi iu}\vartheta_1(u).$$

Show that the functions $\vartheta_0(u)$, $\vartheta_1(u)$ are not identically zero. (Show, for instance, that

$$\int_0^1 |\vartheta_0(x)|^2\,dx = 1 + 2\sum_{n\geqslant 1}|q|^{n^2}.)$$

Show that the complex numbers $m + n\tau$, for integral values of m and n, are zeros of the function $\vartheta_1(u)$, and that the numbers $m+\left(n+\frac{1}{2}\right)\tau$ are zeros of $\vartheta_0(u)$. By evaluating the integral of the function h'/h round the perimeter of a suitable parallelogram of periods, show that there are no other zeros.

4. Let a be a real number. By proceeding as in no. 2, § 2, prove the relation

$$\frac{\pi i \sinh 2\pi a}{\sin\pi(z+ai)\sin\pi(z-ai)} = \sum_{-\infty<n<+\infty}\left(\frac{1}{z+n-ai}-\frac{1}{z+n+ai}\right),$$

and deduce that

$$\frac{\pi}{a}\cdot\frac{\sinh 2\pi a}{\cosh 2\pi a - \cos 2\pi z} = \sum_{-\infty<n<+\infty}\frac{1}{(z+n)^2+a^2}.$$

5. Justify the following expansions :

(i)
$$\frac{\pi}{\cos\pi z} = \sum_{n\geqslant 1}\frac{(-1)^{n+1}(2n-1)}{\left(n-\frac{1}{2}\right)^2 - z^2},$$

(ii)
$$\pi\tan\pi z = 2z\sum_{n\geqslant 1}\frac{1}{\left(n-\frac{1}{2}\right)^2 - z^2}.$$

Deduce from (i) that

$$\frac{\pi}{4} = 1 - \frac{1}{3} + \frac{1}{5} - \frac{1}{7} + \cdots$$

Deduce the following formulae from (i) and (ii) by the methods of no. 3, § 3 :

(iii)
$$\cos\pi z = \prod_{n\geqslant 1}\left(1-\frac{4z^2}{(2n-1)^2}\right),$$

(iv)
$$\cos\frac{\pi z}{4} - \sin\frac{\pi z}{4} = \prod_{n\geqslant 1}\left(\frac{1+(-1)^n z}{2n-1}\right).$$

(Note that

$$\frac{(\cos t/2 - \sin t/2)'}{\cos t/2 - \sin t/2} = -\frac{1}{2}\frac{1+\sin t}{\cos t}.)$$

6. Show that

$$\frac{d}{dz}\left(\frac{\Gamma'(z)}{\Gamma(z)}\right) + \frac{d}{dz}\left(\frac{\Gamma'\left(z+\frac{1}{2}\right)}{\Gamma\left(z+\frac{1}{2}\right)}\right) = 2\frac{d}{dz}\left(\frac{\Gamma'(2z)}{\Gamma(2z)}\right).$$

(Use formula (4. 14)). Deduce, by integration, that

$$\Gamma(z)\,\Gamma\left(z+\frac{1}{2}\right) = e^{az+b}\Gamma(2z), \quad a, b \text{ constants};$$

determine a and b by putting $z = \frac{1}{2}$, 1 in turn. Use the same method to obtain the more general formula for any integer $p \geqslant 2$:

$$\Gamma(pz) = (2\pi)^{-(p-1)/2}p^{pz-1/2}\Gamma(z)\,\Gamma\left(z+\frac{1}{p}\right)\cdots\Gamma\left(z+\frac{p-1}{p}\right).$$

(To determine the constants of integration, put $z = 1/p$, 1; formula (4. 9) with $z = q/p$, $1 \leqslant q \leqslant p$ and the relation $\sin\frac{\pi}{p}\sin\frac{2\pi}{p}\cdots\sin\frac{p-1}{p}\pi = p/2^{p-1}$ (for $p \geqslant 2$) can be used to evaluate

$$\Gamma(1/p)\,\Gamma(2/p)\cdots\Gamma((p-1)/p)).$$

7. (i) Show that the integral

$$\int_0^\infty e^{-t}t^{x-1}\,dt,$$

where x is a real parameter, converges uniformly in any interval $a \leqslant x \leqslant b$ with $0 < a < b$; deduce that the integral $\int_0^\infty e^{-t}t^{z-1}\,dt$ defines a holomorphic function of z, which is denoted by $G(z)$, in the half-plane $\mathrm{Re}(z) > 0$.
(ii) Show that

(1) $$\int_0^n \left(1-\frac{t}{n}\right)^n t^{x-1}\,dt = \frac{n^x n!}{x(x+1)\cdots(x+n)} \quad \text{for real } x > 0$$

and n an integer $\geqslant 1$;

(2) $$e^{-t}\left(1-\frac{e}{2n}t^2\right) \leqslant \left(1-\frac{t}{n}\right)^n \leqslant e^{-t} \quad \text{for } 0 \leqslant t \leqslant n.$$

(First prove the inequalities $1 - t/n \leqslant e^{-t/n} \leqslant 1 - t/n + t^2/2n^2$, and then use the inequality $a^n - b^n \leqslant na^{n-1}(a-b)$, which is true for $a \geqslant b \geqslant 0$, by taking $a = e^{-t/n}$, $b = 1 - t/n$.) Deduce that

$$\lim_n \int_0^n \left(1 - \frac{t}{n}\right)^n t^{x-1}\, dx = \int_0^\infty e^{-t} t^{x-1}\, dx$$

and that

$$G(z) = \Gamma(z) \quad \text{for} \quad \operatorname{Re}(z) > 0$$

8. Determine the residue of the function $\Gamma(z)$ at the pole $z = -n$ for $n = 0, 1, 2, \ldots$.

9. Show that, if

$$\wp(z) = \frac{1}{z^2} + a_2 z^2 + a_4 z^4 + \cdots + a_{2n} z^{2n} + \cdots$$

is the Laurent expansion of the function $\wp(z)$ at the origin, then the differential equation (5. 9) of § 2 allows us to calculate the coefficients a_{2n} for $n \geqslant 3$ as polynomials in a_2 and a_4 by induction. Determine a_6 and a_8.

10. Let P be a parallelogram of periods of the functions $\wp$. Show that, if α and β are two complex numbers, the function

$$\wp'(z) - \alpha\wp(z) - \beta \tag{1}$$

has three zeros in P and that their sum is equal to a period (use propositions 5. 1 and 5. 2 of chapter III, § 5). Deduce that, if u and v are two complex numbers such that $u \pm v \not\equiv 0 \pmod{\Omega}$, then α, β can be found so that the function (1) has zeros at u, v and $-u-v$; deduce that, if $u+v+w=0$, then

$$\det \begin{vmatrix} \wp(u) & \wp'(u) & 1 \\ \wp(v) & \wp'(v) & 1 \\ \wp(w) & \wp'(w) & 1 \end{vmatrix} = 0.$$

11. In the notation of example 3 above, show that the infinite product

$$\prod_{n \geqslant 1} \left[(1 - q^{2n-1} e^{2\pi i u})(1 - q^{2n-1} e^{-2\pi i u})\right]$$

defines a holomorphic function $f(u)$ in the whole of the plane of the complex variable u. What are the zeros of $f(u)$? Show that

$$f(u) = c \,.\, \vartheta_0(u),$$

where c denotes a constant.
(The function $f(u)/\vartheta_0(u)$ is shown to be doubly periodic and holomorphic in the whole plane, so the corollary to proposition 5. 1 of chapter III, § 5 applies.)

CHAPTER VI

Holomorphic Transformations

1. General Theory; Examples

1. Local study of a holomorphic transformation $w = f(z)$ when $f'(z_0) \neq 0$

Proposition 1. 1. *Let $w = f(z)$ be a holomorphic function in a neighbourhood of z_0; suppose that $f'(z_0) \neq 0$, and put $w_0 = f(z_0)$. When z and w are sufficiently near to z_0 and w_0, respectively, the relation $w = f(z)$ is equivalent to a relation $z = g(w)$, where g denotes a (well-defined) holomorphic function of w in some neighbourhood of w_0 such that $g(w_0) = z_0$.*

This follows from chapter I, § 2 proposition 9. 1, and also from chapter IV, § 5, proposition 6. 1.

Hence, in some neighbourhood of a point, the inverse transformation of a holomorphic transformation with derivative $\neq 0$ is a holomorphic transformation; moreover, with the above notation, the derivative g' is given by the relation

$$g'(w) = \frac{1}{f'(z)}.$$

In particular, this derivative is $\neq 0$ at the point w_0.

Let c be the non-zero complex number $f'(z_0)$. The (homogeneous) linear, tangent transformation of the transformation f at the point z_0 is the transformation

$$W = cZ. \tag{1. 1}$$

This, considered as a transformation of the plane, is a *direct similitude.* In particular, this transformation *preserves angles and their orientation.* In other words, if two differentiable paths γ_1 and γ_2 of the z-plane have initial point z_0, the images of these paths under the transformation $w = f(z)$ are differentiable paths with initial point w_0 and the half-tangents at w_0

make the same oriented angle as the half-tangents to the paths γ_1 and γ_2 at the point z_0. For this reason, we say that a holomorphic transformation $w = f(z)$ is *conformal* at each point z_0 where the derivative $f'(z_0)$ is $\neq 0$.

Conversely, any (homogeneous) linear transformation of the plane which preserves angles (without necessarily preserving their orientation) is of the form (1. 1), or of the form

$$\text{(1. 2)} \qquad W = c\overline{Z}.$$

For, if T is such a transformation, there exists a direct similitude S such that the composed transformation $S^{-1} \circ T$ leaves the real point (1,0) fixed. Since $S^{-1} \circ T$ preserves angles, the point (0, 1) is transformed into $(0, a)$, where a is real $\neq 0$. Thus (1, 1) is transformed into $(1, a)$, the vectors (1, 1) and $(1, a)$ make equal angles with (1, 0), whence $a = \pm 1$. If $a = 1$, $S^{-1} \circ T$ is the identity and $T = S$ is of the form (1. 1). If $a = -1$, $S^{-1} \circ T = U$ is a symmetry with respect to the real axis and $T = S \circ U$ is of the form (1. 2), which completes the proof.

In the case (1. 1), the linear transformation preserves orientation; in the case (1. 2), it changes orientation. Consider, then, a transformation $w = f(z)$ defined in a *connected* open set D of the plane of the complex variable $z = x + iy$; suppose it to be continuously differentiable with Jacobian $\neq 0$ at all points of D; if this transformation preserves angles (in other words, if the linear tangent transformation is one of the types (1. 1) or (1. 2), at each point of D), then, at each point of D, one of the relations

$$\frac{\partial f}{\partial \bar{z}} = 0, \qquad \frac{\partial f}{\partial z} = 0,$$

holds. These relations can never be satisfied simultaneously at a point of D because the partial derivatives of f with respect to x and y would be both zero, contradicting the fact that the Jacobian is non-zero. Since the two functions $\frac{\partial f}{\partial \bar{z}}$ and $\frac{\partial f}{\partial z}$ are continuous, the sets of points of D on which they are zero are closed in D; hence, D is the union of these two disjoint closed sets, and, consequently, one of the two sets is empty since D is connected. Only two cases are possible : either $\frac{\partial f}{\partial \bar{z}} = 0$ at any point of D (so the transformation is holomorphic), or $\frac{\partial f}{\partial z} = 0$ at any point of D (so f is a holomorphic function of $\bar{z}$). In the latter case we say that the transformation is *antiholomorphic*. To sum up :

PROPOSITION 1. 2. *A necessary and sufficient condition for a continuously differentiable transformation with Jacobian* $\neq 0$ *everywhere in a connected open set* D *of the plane to preserve angles, is that it is either holomorphic or antiholomorphic. In the first case, it conserves the orientation of angles and, in the second case, it changes the orientation of angles.*

2. Local study of a holomorphic transformation $w = f(z)$ when $f'(z_0) = 0$.

First, consider a particular case, that of the transformation

$$w = z^p, \tag{2.1}$$

where p denotes an integer $\geqslant 2$. The derivative of z^p is zero for $z = 0$. The inverse transformation

$$z = w^{1/p}, \tag{2.2}$$

is many-valued : to each value $\neq 0$ of w, there correspond p distinct values of z. Angles at the origin are not preserved by the transformation (2. 1) because the argument of w is p times the argument of z. The angles are, in fact, multiplied by the integer p. When the point z turns once round the origin, the point w turns p times round the origin in the same direction; the reader is invited to formulate this precisely in terms of the *index* of any closed curve described by z and the index of the transformed curve described by w.

To study the general case of a holomorphic transformation $w = f(z)$ when $f'(z_0) = 0$, we suppose for that $z_0 = 0$, $f(z_0) = 0$ for simplicity. In what follows, it is essential to assume that f is not identically zero in a neighbourhood of 0; if p is the order of multiplicity of the zero of f at the origin, the Taylor expansion of f at the origin is of the form

$$w = cz^p(1 + f_1(z)), \tag{2.3}$$

where the constant c is $\neq 0$ and the function f_1, which is holomorphic at the origin, satisfies $f_1(0) = 0$. Put

$$f_2(z) = c^{1/p}(1 + f_1)^{1/p};$$

the function $f_2(z)$ is holomorphic in some neighbourhood of the origin (we choose one of its branches), and $f_2(0) \neq 0$. Relation (2. 3) then becomes

$$w = (zf_2(z))^p. \tag{2.4}$$

Let

$$zf_2(z) = t. \tag{2.5}$$

By no. 1, this relation gives $z = g(t)$ where g is holomorphic in some neighbourhood of 0 and zero at the point 0 with $g'(0) \neq 0$. By (2. 4), we have $t = w^{1/p}$ whence, finally,

$$z = g(w^{1/p}). \tag{2.6}$$

Hence, the relation $w = f(z)$ *is equivalent, in a neighbourhood of the origin, to a relation of the form* (2. 6), *where* g *is holomorphic in a neighbourhood of* 0 *and zero at the origin, with* $g'(0) \neq 0$.

In particular, to any value of w sufficiently near to 0 and $\neq 0$ there correspond p distinct values of z. The origin is said to be a *critical point* of order p for the transformation (2.6), the inverse of $w = f(z)$.

3. Holomorphic transformations

Theorem. *Let* f *be a holomorphic function, which is not constant, in a connected open set* D. *Then the image* $f(\mathrm{D})$ *is an open set of the plane.*

Proof. It is sufficient to show that, for any point $z_0 \in \mathrm{D}$, the image $f(\mathrm{D})$ contains all the points of some neighbourhood of $f(z_0)$. The case when $f'(z_0) \neq 0$ follows from no. 1 : in this case f defines a homeomorphism of a neighbourhood of z_0 onto a neighbourhood of $f(z_0)$. The case when $f'(z_0) = 0$ (the function f not being constant in any neighbourhood of z_0) follows from no. 2 : in this case there is a neighbourhood of z_0 in which the function f takes p times. each value sufficiently near to $f(z_0)$ and $\neq f(z_0)$. Hence, the theorem is proved in all cases.

Note. For any open subset D′ of D, the image $f(\mathrm{D}')$ is open. We say then that f is an *open mapping.*

Corollary. *If* f *is a simple* (cf. chapter, v, § 1, no. 2) *holomorphic function in a connected open set* D, *then* f *is a homeomorphism of* D *onto the open set* $f(\mathrm{D})$, *and the inverse mapping* f^{-1} *is holomorphic in* $f(\mathrm{D})$.

Proof. f is an injective, continuous, open mapping. Its inverse mapping f^{-1} is continuous because f is open. Since f is simple, $f'(z_0) \neq 0$ at any point $z_0 \in \mathrm{D}$ by the results of no. 2; thus, by no. 1, f^{-1} is holomorphic at each point $f(z_0)$.

Definition. Let D be an open set of the plane of the variable z, and D′ an open set of the plane of the variable w. An *isomorphism* of D on D′ is defined to be a homeomorphism which is defined by a *holomorphic* mapping f whose inverse is also holomorphic.

It follows from the above corollary that if a holomorphic mapping of D is simple, then it is an *isomorphism* of D onto its image $f(\mathrm{D})$.

Note. The above definitions and results hold, not only when D is an open set in the plane of a complex variable, but, more generally, when D is an open set of the Riemann sphere, and the mapping f can also take values in the Riemann sphere.

4. Examples of non-simple holomorphic functions

Even when the derivative $f'(z)$ is everywhere $\neq 0$, the function f can be non-simple (that is to say not simple). The easiest example is the transformation

$$w = e^z$$

which is periodic of period $2\pi i$. A strip $a < \operatorname{Im} z < b$ is transformed into the set of points w such that

$$a < \arg w < b.$$

In this strip, the transformation is simple if and only if

$$b - a \leqslant 2\pi.$$

By way of an example, we shall study the transformation $w = \cos z$ whose derivative vanishes at all z which are integral multiples of π. We have

$$w = \cos z = \frac{1}{2}(e^{iz} + e^{-iz}),$$

so the transformation $w = \cos z$ is composed of two transformations

$$t = e^{iz} \quad \text{and} \quad w = \frac{1}{2}(t + 1/t).$$

Let us examine the inverse transformation : if w is given arbitrarily, two values of t correspond to it; they are the roots of the quadratic equation

$$t^2 - 2wt + 1 = 0,$$

and their product is therefore equal to 1; they are distinct if $w \neq \pm 1$; to each of these roots, there corresponds an infinity of values of z deduced from each other by adding arbitrary integral multiples of 2π.

Put $z = x + iy$, $w = u + iv$ (x, y, u, v being real). Then

$$u = \cosh y \cos x, \qquad v = -\sinh y \sin x.$$

If y is kept and x varies, the point (u, v) describes the ellipse

$$\frac{u^2}{\cosh^2 y} + \frac{v^2}{\sinh^2 y} = 1$$

(the point describes the ellipse once each time that x describes an interval of length 2π). If x is kept fixed while y varies, the point (u, v) describes once one of the two branches of the hyperbola

$$\frac{u^2}{\cos^2 x} - \frac{v^2}{\sin^2 x} = 1.$$

To study how w varies as a function of z, it is sufficient, because of the periodicity, to let x vary from $-\pi$ to $+\pi$ and y from $-\infty$ to $+\infty$ Moreover, if we put $-z$ for z, w remains unchanged, so we shall only let x

vary from 0 to π. If we put $-y$ for y without changing x, u remains unchanged, but v is changed into $-v$: thus, to a pair of points z_1, z_2 which are symmetric about the real axis, there corresponds a pair of points w_1, w_2 which are symmetric about the real axis. Finally, it is sufficient to let x vary from 0 to π and y from 0 to $+\infty$. Let D_1 then, be the open set

$$(D) \qquad 0 < x < \pi, \qquad y > 0.$$

First, let $z = x + iy$ describe the oriented boundary of D: 1° while x stays at 0 and y decreases from $+\infty$ to 0, w stays on the real axis decreasing from $+\infty$ to $+1$; 2° while y stays at 0 and x increases

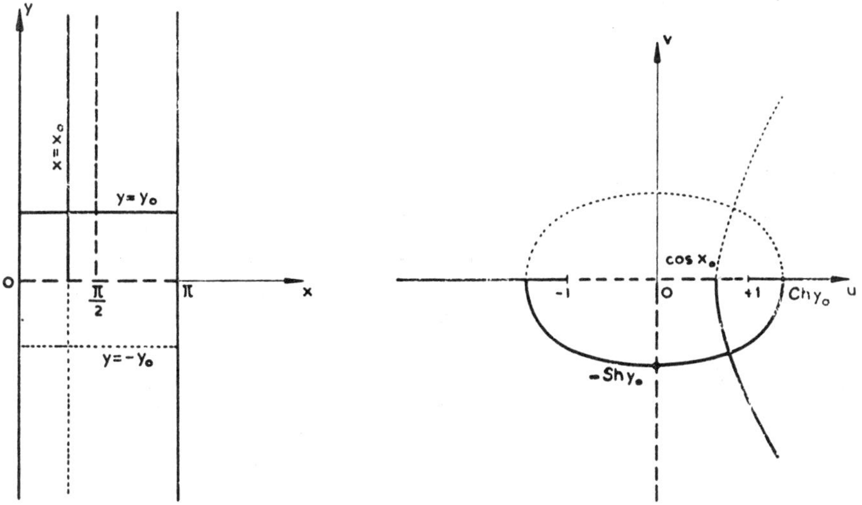

Fig. 11

from 0 to π, w stays on the real axis decreasing from $+1$ to -1; 3° when x stays equal to π, y increases from 0 to $+\infty$, w stays on the real axis decreasing from -1 to $-\infty$. Hence, the mapping $w = \cos z$ maps the boundary of D homeomorphically onto the real axis.

The reader should prove that D is mapped homeomorphically onto the lower half-plane $v < 0$. In fact, when the point z describes a segment $y = y_0$ (constant > 0) as x increases from 0 to π, the point w decribes, exactly once, the half-ellipse which is situated in the half-plane $v < 0$ with $+1$ and -1 as foci and whose semi-major and semi-minor axes are $\cosh y_0$ and $\sinh y_0$, respectively. When the point z_0 describes a half-line $x = x_0$ $(0 < x_0 < \pi)$ and y increases from 0 to $+\infty$, the point w describes, just once, the half-branch of the hyperbola which is situated in the half-plane $v < 0$ with $+1$ and -1 as foci and whose semi-axes are $|\cos x_0|$ and $\sin x_0$.

The strip $0 < x < \pi$ (y varying from $-\infty$ to $+\infty$) is mapped homeomorphically onto the plane excluding the two half-lines of the real axis. $u \geqslant +1$ and $u \leqslant -1$.

Let us examine the angles. We note that the transformation $w = \cos z$ doubles the angles at each of the points $z = 0$ and $z = \pi$ (the right angles in the boundary of D become flat); this corresponds to the fact that derivative $-\sin z$ of $\cos z$ has a simple zero at each of these points. Angles are preserved by the transformation in the interior of D; lines parallel to the coordinate axes in the z-plane are transformed into a confocal system of ellipses and hyperbolae as we have seen.

2. Conformal Representation

1. Statement of the problem

Let D and D′ be two connected open sets of the Riemann sphere $\mathbf{S}_2$. We ask whether there exists an isomorphism of D on D′, or, what is equivalent, whether there is a simple holomorphic mapping of D on D′. A necessary condition for a positive answer to this problem is a purely topological one : the sets D and D′ *must* be homeomorphic, for any isomorphism is a homeomorphism. For example, if D is simply connected, it is necessary that D′ is also simply connected. This necessary condition is not sufficient as the following theorem shows :

THEOREM 1. *The plane* $\mathbf{C}$ *and the open disc* $|z| < 1$ *are not isomorphic* (although they are homeomorphic).

Proof. Suppose that there is an isomorphism f of $\mathbf{C}$ on the disc $|z| < 1$. Then, f is a bounded holomorphic function, which must be constant by Liouville's theorem, and this contradicts the fact that f is simple.

2. Automorphisms of D

Let us suppose that there is at least one isomorphism f of D on D′ and that we want to find *all the isomorphisms* g of D on D′. The transformation $S = f^{-1} \circ g$ is an isomorphism of D onto itself, in other words, it is an automorphism of D; thus,

$$g = f \circ S. \tag{2.1}$$

Conversely, if S is an automorphism of D, the transformation g defined by (2. 1) is an isomorphism of D on D'. Hence, we obtain all the isomorphisms of D on D' by composing an arbitrary automorphism of D with a particular isomorphism f of D on D'. It is clear that the automorphisms of D form a *group* $\Gamma(D)$. Moreover, if f is an isomorphism of D on D', the mapping $S \to f \circ S \circ f^{-1}$ is an isomorphism of the group $\Gamma(D)$ on the group $\Gamma(D')$.

We propose to determine the group $\Gamma(D)$ explicitly for certain simple open sets in the examples which follow.

3. Automorphisms of the complex plane

Here we take the case when D is the whole complex plane **C**. Let $z \to f(z)$ be an automorphism of **C**; the function $f(z)$ is holomorphic in **C**; so only two possible cases arise *a priori* :

1° f has an essential singularity at the point at infinity;

2° f is a polynomial.

We shall see that case 1° is impossible. For, since f is simple, the image under f of the annulus $|z| > 1$ does not meet the image under f of the disc $|z| < 1$; this image is open and non-empty. Thus the image of $|z| > 1$ is not dense in the whole plane, and, consequently, by the Weierstrass theorem (chapter III, § 4, no. 4) the point at infinity is not an essential singularity of f. Hence, f is a polynomial of degree $n \geqslant 1$, and, by d'Alembert's theorem, the equation $f(z) = w$ has n distinct roots (except for particular values of w). However, f is simple by hypothesis, so we conclude that $n = 1$. We have thus proved the following theorem :

THEOREM 2. *The automorphism group of* **C** *consists of the linear transformations*

$$z \to az + b, \qquad a \neq 0. \tag{3. 1}$$

When $a = 1$, the transformation (3. 1) is a *translation*; it has no fixed points. On the other hand, when $a \neq 1$, the transformation has a unique fixed point

$$z = \frac{b}{1 - a}.$$

We note that the transformations (3. 1) form a *transitive* group in the plane **C** : in other words, given any pair of points z_1, z_2, there is at least one transformation of the group which takes z_1 into z_2. The *stabilizer* of a point z_0, that is the subgroup of transformations which leave the point z_0 fixed, is easily found; for example, the stabilizer of the origin 0 is composed of the direct similitudes $z \to az$.

4. Automorphisms of the Riemann sphere

Consider the homographic transformations

$$w = \frac{az + b}{cz + d}, \qquad ad - bc \neq 0. \tag{4.1}$$

If the constants a, b, c, d, are multiplied by the same complex number $\neq 0$, the transformation remains the same. Hence, we must always consider that the coefficients a, b, c, d, are only defined up to a constant factor.

Such a transformation is defined on the Riemann sphere $\mathbf{S}_2$ and takes values in the Riemann sphere $\mathbf{S}_2$: to be precise, for $z = \infty$, $w = a/c$ if $c \neq 0$, and $w = \infty$ if $c = 0$ (which implies that $a \neq 0$). Each transformation (4. 1) has an inverse transformation

$$z = \frac{dw - b}{-cw + a}, \tag{4.2}$$

which shows that each homographic transformation (4. 1) is a homeomorphism of $\mathbf{S}_2$ on $\mathbf{S}_2$.

The transformations (4. 1) thus form a group G of automorphisms of the Riemann sphere $\mathbf{S}_2$. We propose to prove the following :

Theorem 3. *The Riemann sphere has no other automorphism than the homographies* (4. 1).

Proof. Consider the subgroup of G whose elements are the transformations leaving the point at infinity of $\mathbf{S}_2$ fixed. They are the transformations with $c = 0$, and, since $d \neq 0$, we can suppose $d = 1$. In other words, the subgroup of G leaving the point at infinity fixed is merely the group of all the automorphisms $w = az + b$ of the plane $\mathbf{C}$ (theorem 2). This group is then the group of all the automorphisms of $\mathbf{S}_2$ which leave the point at infinity fixed. Theorem 3 thus follows from a lemma of a general nature :

Lemma. *Let* D *be an open set of the Riemann sphere* $\mathbf{S}_2$ *and let* G *be a subgroup of the group* $\Gamma(\mathrm{D})$ *of all automorphisms of* D. *Suppose that the following conditions are satisfied* :

a) G *is transitive in* D;

b) *there is at least one point of* D *whose stabiliser is contained in* G.

Then, G *is the group of all automorphisms of* D.

Proof of the lemma : Let $\mathrm{S} \in \Gamma(\mathrm{D})$, and let $z_0 \in \mathrm{D}$ be a point whose stabilizer is contained in G. Since G is transitive there is a $\mathrm{T} \in \mathrm{G}$ such that

$T(z_0) = S(z_0)$. Hence, the transformation $T^{-1} \circ S \in \Gamma(D)$ leaves the point z_0 fixed and therefore belongs to G; thus $S = T \circ (T^{-1} \circ S)$ belongs to G.

This completes the proof.

5. Geometrical study of the group of homographies; equivalence of the half-plane to the disc.

When $c \neq 0$, transformation (4. 1) can be put in the well-known canonical form

$$w = \frac{a}{c} + \frac{(bc - ad)/c^2}{z + d/c}. \tag{5. 1}$$

It follows that (4. 1) is composed of the transformations

$$z_1 = z + \frac{d}{c}, \quad z_2 = \frac{1}{z_1}, \quad z_3 = kz_2, \quad w = z_3 + \frac{a}{c},$$

where $k = \dfrac{bc - ad}{c^2}$ and where each is a special type of homography. Hence, any homography is composed of translations, homotheties with factor $\neq 0$, and inversion-reflection (an inversion-reflection is a transformation of the form $z' = 1/z$; such a transformation is composed of a reflection about the real axis and an inversion of centre 0 and radius 1). This result has been established for transformations (4. 1) such that $c \neq 0$; when $c = 0$, it is still true in an obvious way. We deduce that any homographic transformation takes circles, or lines, into circles, or lines (lines can be considered as circles passing through the point at infinity). Moreover, homographic transformations are conformal since they are holomorphic mappings of $\mathbf{S}_2$ into $\mathbf{S}_2$; in particular, they transform orthogonal circles (or lines) into orthogonal circles (or lines).

For an arbitrarily given pair of circles (or lines), there is always a homography transforming one into the other. In particular, there is a homoaraphy which takes the real axis $y = 0$ into the unit circle : an example is the transformation

$$w = \frac{z - i}{z + i}. \tag{5. 2}$$

To verify this, it is sufficient to check that three particular points of the real axis (for example 0,1 and ∞) are transformed into points of the unit circle (in this case, the points $w = -1$, $w = -i$ and $w = 1$).

A priori, a homographic transformation which takes the real axis into the unit circle transforms one of the half-planes bounded by the real axis into the interior of the unit disc, and the other half-plane into the exterior of the unit disc (including the point at infinity). In the case of transformation (5. 2), the upper half-plane $y>0$ is transformed into the disc $|w|<1$, since the point $z=i$ is transformed into $w=0$.

6. Automorphisms of the half-plane and of the unit disc

Let P denote the upper half-plane $y>0$ and let B denote the open disc $|w|<1$. By the results at the end of no. 2, the transformation (5. 2) establishes an isomorphism of the group $\Gamma(\mathrm{P})$ onto the group $\Gamma(\mathrm{B})$. We propose now to determine these groups explicitly.

We have already determined the group of all automorphisms of the Riemann sphere. A subgroup of these is formed by the ones which transform the real axis $y=0$ into itself; it is the subgroup consisting of the homographic transformations

$$z \to \frac{az+b}{cz+d}, \qquad ad-bc \neq 0, \tag{6. 1}$$

with *real* coefficients a,b,c,d. For, it is obvious that, if the coefficients are real, the transformations (6. 1) take the real axis into itself; conversely, if the real axis is transformed into itself, the coefficients a,b,c,d are determined up to a factor by a system of linear equations with real coefficients, which are obtained by considering three distinct points z_1,z_2,z_3 of the real axis and expressing that their transforms are real.

Since the cofficients of (6. 1) are only defined up to a real factor $\neq 0$, we can suppose that $ad-bc=\pm 1$ in (6. 1). It is easy to see that the transformations of the form (6. 1) which take the upper half-plane $y>0$ into itself are those for which $ad-bc=1$; for, it is sufficient to verify that the imaginary part of $\dfrac{ai+b}{ci+d}$ is >0. The transformations (6. 1) for which $ad-bc=1$ form a subgroup G of the group $\Gamma(\mathrm{P})$ of all automorphisms of the half-plane P; each transformation of G determines the coefficients a, b, c, d up to a factor ± 1.

Theorem 4. *The above group* G *contains all the automorphisms of the half-plane* P.

When we have proved this theorem, it will follow that every automorphism of the half-plane P extends to an automorphism of the Riemann sphere, which is not obvious *a priori*.

To show that $\mathrm{G}=\Gamma(\mathrm{P})$, we observe first that the group G is transitive in the half-plane P. For, the point i can be transformed into an arbitrary point $a+ib$ $(b>0)$ of the half-plane by a suitable transformations in G; this follows immediately. If we show that the stabilizer of a point in the half-plane (for example, the point $z = i$) is contained in G, then theorem 4 will be proved because of the lemma of no. 4. The proof is then reduced to showing that the stability subgroup of the point i consists of homographic transformations. Transformation (5. 2) defines an isomorphism of this stability subgroup onto the subgroup $|w| < 1$ of $\Gamma(B)$ consisting of the automorphisms of the disc which leave the centre 0 fixed. It is therefore sufficient to prove the following :

PROPOSITION 6. 1. *An automorphism of the disc $|z| < 1$, which leaves 0 fixed, is a rotation $z \to ze^{i\theta}$ for some angle θ.*

Proof of proposition 6. 1. Let $z \to f(z)$ be an automorphism of the unit disc such that $f(0) = 0$. By Schwarz' lemma (chapter III, § 3), we have

$$|f(z)| \leqslant |z|$$

for all z such that $|z| < 1$. But, by applying Schwarz' lemma to the inverse transformation, we also find that

$$|z| \leqslant |f(z)|.$$

Comparing them gives $|z| = |f(z)|$, and consequently, again by Schwarz lemma, we have $f(z) = cz$, where c is a constant of unit modulus. This completes the proof.

Hence, we have also completed the proof of theorem 4.

As an exercise one can determine explicitly the stabilizer of the point $z = i$ in the group of all automorphisms of the upper half-plane $y > 0$. It is transformed from the stabilizer of 0 into the group of automorphisms of the unit disc by the transformation (5. 2). One obtains the transformations

$$z \to \frac{z + \tan \frac{\theta}{2}}{1 - z \tan \frac{\theta}{2}}$$

which depend on the real parameter θ.

To determine the automorphism group of the unit disc $|z| < 1$, it is sufficient to transform the group of automorphisms of the upper half-plane by (5. 2). However, we shall use a more direct method. The problem is to find all the homographic transformations

$$z' = \frac{az + b}{cz + d}$$

which transform the circle $z\bar{z} - 1 = 0$ into the circle $z'\bar{z}' - 1 = 0$ while transforming the open disc $1 - z\bar{z} > 0$ into the open disc $1 - z'\bar{z}' > 0$. The first of these conditions says that

$$(az + b)(\bar{a}\bar{z} + \bar{b}) = (cz + d)(\bar{c}\bar{z} + \bar{d})$$

for all z of modulus 1, which implies

$$a\bar{b} = c\bar{d} \tag{6.2}$$

and

$$a\bar{a} - c\bar{c} = d\bar{d} - b\bar{b}. \tag{6.3}$$

We have then

$$1 - z'\bar{z}' = \frac{(d\bar{d} - b\bar{b})(1 - z\bar{z})}{|cz + d|^2},$$

and, for $1 - z\bar{z} > 0$ to imply $1 - z'\bar{z}' > 0$, it is necessary and sufficient that

$$d\bar{d} - b\bar{b} > 0. \tag{6.4}$$

Inequality (6.4) along with equation (6.3) implies that $a \neq 0$, $d \neq 0$; from (6.2), we have

$$\frac{c}{a} = \frac{\bar{b}}{\bar{d}} = \bar{\lambda}, \quad \text{with} \quad |\lambda| < 1,$$

and, by (6.3),

$$|a| = |d|.$$

Hence,

$$\frac{az + b}{cz + d} = \frac{a}{d} \frac{z + \lambda \frac{d}{a}}{1 + \bar{\lambda} \frac{a}{d} z} = e^{i\theta} \frac{z + z_0}{1 + \bar{z}_0 z},$$

where θ is real and z_0 is a complex number such that $|z_0| < 1$.

To sum up, we have proved the following :

PROPOSITION 6.2. *The group of automorphisms of the unit disc is composed of the homographic transformations of the form*

$$z' = e^{i\theta} \frac{z + z_0}{1 + \bar{z}_0 z}, \quad \text{for } \theta \text{ real and } |z_0| < 1. \tag{6.5}$$

3. Fundamental Theorem of Conformal Representation

1. STATEMENT OF THE FUNDAMENTAL THEOREM

We propose the following problem : given an open set D of the plane **C**, find all the isomorphisms (if any exist) of D on the unit disc $|z|$ 1. A< necessary condition for the existence of such an isomorphism is the following :

It is necessary that D *shall be simply connected and different from* **C**.

The first part of the condition is necessary because D must be homeomorphic to the open disc, which is simply connected; the condition $D \neq \mathbf{C}$ is necessary because of theorem 1 of § 2, no. 1. The following fundamental theorem states that these necessary conditions are also sufficient.

FUNDAMENTAL THEOREM. *Any open set* D *of the plane* **C** *which is simply connected and different from* **C** *is isomorphic to the open disc* $|z| < 1$.

The proof of this theorem will be the object of numbers 3 and 4. First of all, we observe that any isomorphism of D onto $|z| < 1$ is composed of a particular isomorphism and an arbitrary automorphism of the unit disc. Since the automorphisms of the unit disc form a transitive group, we see that, if there exists an isomorphism of D on the unit disc, there exists an isomorphism which takes an arbitrarily chosen point $z_0 \in D$ into the centre of the disc. Hence, we restrict the required isomorphism f with condition

$$f(z_0) = 0. \tag{1.1}$$

Moreover, the stabilizer of the centre of the unit disc consists of the rotations about 0 (§ 2, proposition 6. 1); we can thus restrict the isomorphism f with the additional condition

$$f'(z_0) \text{ is real and } f'(z_0) > 0. \tag{1.2}$$

In brief, conditions (1. 1) and (1. 2) completely determine the required isomorphism f if it exists.

We shall immediately state two corollaries of the fundamental theorem.

COROLLARY 1. *Two simply-connected open sets* D_1 *and* D_2 *of the plane* **C** *are isomorphic if they are both different from* **C**.

We note that, because of theorem 1 of § 2, no. 1, a simply-connected open set D different from **C** is not isomorphic ot **C**. Nevertheless :

COROLLARY 2. *Two simply connected open sets* D_1 *and* D_2 *of the plane are always homeomorphic.*

For, if they are different from **C**, this follows from corollary 1; if one of them is equal to **C**, it follows from the fact that the disc $|z| < 1$ is *homeomorphic* to the plane **C**.

2. REDUCTION TO THE CASE OF A BOUNDED DOMAIN

PROPOSITION 2. 1. *Let* D *be an open set satisfying the hypotheses of the fundamental theorem. Then, there exists an isomorphism of* D *onto a bounded open set of the plane* **C**.

For, there exists a point $a \notin D$ by hypothesis. Consider the function $\log(z - a)$ in D; a branch $g(z)$ of it can be chosen since D is simply-connected (cf. chapter II, § 1, no. 7). The holomorphic function g is simple in D because the relation $g(z_1) = g(z_2)$ implies

$$e^{g(z_1)} = e^{g(z_2)} \qquad \text{that is} \qquad z_1 - a = z_2 - a.$$

Choose a point $z_0 \in D$; the function g takes all the values in a disc E of centre $g(z_0)$ as z varies in D (cf. § 1, n°. 1). If we translate this disc by a translation $2\pi i$, we obtain a disc, which has no points in common with the image of D by g since the function e^g is simple. It follows that the function

$$\frac{1}{g(z) - g(z_0) - 2\pi i}$$

is holomorphic, simple and *bounded* in D. It therefore defines an isomorphism of the open set D onto a bounded open set of the plane $\mathbf{C}$, and proposition 2. 1 is proved.

In future, we shall assume D to be *bounded*; by means of a translation and a homothety we can suppose that $z_0 = 0$ and that D is contained in the disc $|z| < 1$. These hypotheses will always be made in future.

3. An extremal property

Proposition 3. 1. *Let* A *be the set of simple holomorphic functions in* D *which satisfy the two conditions*

$$(3.1) \qquad f(0) = 0, \quad |f(z)| < 1 \qquad \text{for} \qquad z \in D.$$

A necessary and sufficient condition for the image D′ *of* f *to be exactly the unit disc is that* $|f'(0)|$ *is maximum in the set of values which it takes as* f *describes* A.

Proof.

1° The condition is necessary. Let $f \in A$, let D′ be its image and let g be an isomorphism of D on the unit disc such that $g(0) = 0$. Then, $f = h \circ g$, where h is an isomorphism of the unit disc onto the image D′ of f, with $h(0) = 0$. Thus, $|h'(0)| \leqslant 1$ by Cauchy's inequality. Hence

$$|f'(0)| \leqslant |g'(0)|.$$

2° The condition is sufficient. To see this, we shall show that if $f \in A$ and if there is an a (with $|a| < 1$) which does not belong to the image of f, then there exists a $g \in A$ such that

$$|g'(0)| > |f'(0)|.$$

For, let us consider the function

$$F(z) = \log \frac{f(z) - a}{1 - \bar{a} f(z)}; \tag{3. 2}$$

it is holomorphic and simple in D. The values of $f(z)$ are in the unit disc, the values of $\frac{f(z) - a}{1 - \bar{a} f(z)}$ are also in the unit disc (cf. § 2, proposition 6. 2), and, consequently, the function $F(z)$ has real part < 0. We have, of course, chosen some branch of the logarithm for $F(z)$, which is possible because D is simply connected. Consider the function

$$g(z) = \frac{F(z) - F(0)}{F(z) + \overline{F(0)}}, \tag{3. 3}$$

which is holomorphic and simple in D. Then, $g(0) = 0$; moreover, $|g(z)| < 1$ because of the following lemma :

LEMMA. *If two complex numbers u and v satisfy* $\operatorname{Re}(u) < 0$ *and* $\operatorname{Re}(v) < 0$, *then* $\left|\frac{v - u}{v + \bar{u}}\right| < 1$. (The proof is left to the reader.)

Hence, the function g belongs to the set A defined in proposition 3. 1. Let us calculate the derivative of g at the origin :

$$g'(0) = \frac{F'(0)}{F(0) + \overline{F(0)}}, \quad \text{with} \quad F'(0) = \left(\bar{a} - \frac{1}{a}\right) f'(0). \tag{3. 4}$$

We then have

$$\frac{|g'(0)|}{|f'(0)|} = \frac{1 - a\bar{a}}{2|a| \log \left|\frac{1}{a}\right|}, \tag{3. 5}$$

and, to show that $|g'(0)| > |f'(0)|$, it is sufficient to verify the inequality

$$\frac{1 - t^2}{t} - 2 \log \frac{1}{t} > 0 \quad \text{for} \quad 0 < t < 1. \tag{3. 6}$$

This verification is elementary : the left hand side is a function of t whose derivative is < 0; thus it is strictly decreasing in the interval $0 < t \leqslant 1$ and, since it is equal to 0 for $t = 1$, it is > 0 for $0 < t < 1$.

We have thus completed the proof of proposition 3. 1.

4. PROOF OF THE FUNDAMENTAL THEOREM

Because of proposition 3. 1, we need only prove that there exists a function $f \in A$ for which the upper bound of $|f'(0)|$ is attained.

Let B be the set of those $f \in A$ for which $|f'(0)| \geqslant 1$. The set B is *non-empty* because the function $f(z) = z$ belongs to it. The set B is a *bounded* subset of the vector space $\mathcal{H}(D)$ (cf. chapter V, § 4, no. 1); for, we have $|f(z)| < 1$ for all $z \in D$ and all $f \in B$. We shall show that B is also a *closed* subset of $\mathcal{H}(D)$. Let f be a holomorphic function in D which is the limit (uniform on compact subsets of D) of a sequence of functions $f_n \in B$. Then,

$$f(0) = \lim f_n(0) = 0.$$

Moreover, since the derivative f' is the limit, uniform on compact sets, of the derivatives f'_n, we deduce that the limit $|f'(0)| = \lim_n |f'_n(0)| \geqslant 1$. Thus the function f is not constant in D. Moreover f is the limit of simple functions f_n. It follows that f is simple (cf. chapter v, § 1, proposition 2. 2). Since $|f_n(z)| < 1$ for all $z \in D$, we have $|f(z)| \leqslant 1$ in the limit; but $|f(z)| = 1$ is impossible at any point $z \in D$ because of the maximum modulus principle, remembering again that f is not constant. To sum up, we have just proved that the function f satisfies all the conditions which define the set B. In other words, $f \in B$, and this proves that the set B is closed in $\mathcal{H}(D)$.

Hence the set B is a bounded, closed subset of $\mathcal{H}(D)$. By the fundamental theorem of chapter v (§ 4, no. 2), the set B is therefore *compact*. However the mapping which associates the real number $|f'(0)|$ to each $f \in B$ is a continuous mapping by theorem 2 of chapter v, § 1, no. 2. This continuous function on a compact space thus attains its upper bound and the fundamental theorem is proved.

4. Concept of complex Manifold; Integration of Differential Forms

1. Complex manifold structure

Let X be a *Hausdorff* topological space. We suppose that an open covering $(U_i)_{i \in I}$ is given, and that for each U_i there is a complex-valued function z_i defined in U_i which is a homeomorphism of U_i onto an open set A_i of the plane **C**. The following coherence condition is imposed on the functions :

(1. 1) *For any $i \in I$ and $j \in I$ such that $U_i \cap U_j \neq \emptyset$, the mapping $f_{ij} = z_i \circ (z_j)^{-1}$ of the image $z_j(U_i \cap U_j) \subset A_j$ onto the image $z_i(U_i \cap U_j) \subset A_i$ is a holomorphic transformation whose derivative is everywhere $\neq 0$.* In other words, $z_i = f_{ij}(z_j)$ in $U_i \cap U_j$, where the f_{ij} are holomorphic (with derivative $\neq 0$) in the open set $z_j(U_i \cap U_j)$ of **C**.

By definition, the datum on X of an open covering and functions z_i satisfying condition (1. 1), is called a *complex manifold structure* (of complex dimension 1). The function z_i is called the *local coordinate* in the open set U_i. If a point of X belongs to several open sets U_i, then there are several local coordinates in some neighbourhood of this point (one for each open set); the change from one local coordinate z_j to another z_i is made by a *holomorphic* transformation f_{ij} because of the coherence condition (1. 1).

For example we have already defined the *Riemann sphere* as a datum of this kind (chapter III, § 5, no. 1). In that case, X was the unit sphere in $\mathbf{R}^3$, and the covering consisted of two open sets, each being the complement of one of the poles of the sphere.

Definition of a holomorphic function. Let X be a space with an complex manifold structure as above. Let f be a continuous complex-valued function defined on X, and, for each i, let f_i be the function defined on the open set $A_i \subset \mathbf{C}$ by the equation $f = f_i \circ z_i$ in U_i. We say that f is *holomorphic* if the function f_i is holomorphic in the open set A_i for each i. In other words, f is holomorphic in X if it can be expressed as a holomorphic function of the local coordinate z_i in each open set U_i.

2. Holomorphic mappings; induced structure

Definition. Let X and Y be two spaces, each with an complex manifold structure. We say that a mapping $\varphi : X \to Y$ is a *holomorphic* mapping if it is continuous and if, in addition, it satisfies the following condition: for any point $a \in X$, let $b = \varphi(a)$ and let w be a local coordinate in a neighbourhood of b in the space Y; then, $w \circ \varphi$ must be a holomorphic function in some neighbourhood of the point a of X. This condition means that $w \circ \varphi$ is expressible as a holomorphic function of a local coordinate in a neighbourhood of a in the space X. Hence, the continuous mapping φ is holomorphic if, in a neighbourhood of each point $a \in X$, a local coordinate in a neighbourhood of the image point $b = \varphi(a)$ is a holomorphic function of a local coordinate in a neighbourhood of a. The coherence condition (1. 1) ensures that the above condition is independent of the choice of local coordinates.

Let X, Y and Z be three complex manifolds and let $\varphi : X \to Y$ and $\psi : Y \to Z$ be two holomorphic mappings. Then, the composed mapping $\psi \circ \varphi$ is a holomorphic mapping of X into Z. The proof is left to the reader.

Let X and Y be two complex manifolds; an *isomorphism* of X on Y is defined to be a homeomorphism $\varphi : X \to Y$ which is holomorphic along with its inverse homeomorphism φ^{-1}. In fact, we shall see later (proposition 6. 1) that, if φ is *a holomorphic homeomorphism*, its inverse mapping is automatically holomorphic, and, consequently, φ is an isomorphism.

Consider two complex manifold structures on the same topological space X, the first being defined by an open cover (U_i) and local coordinates z_i, and the second being defined by an open cover (V_α) and local coordinates w_α. We ask if the identity mapping $X \to X$ is an isomorphism of the first complex manifold structure onto the second. By going back to the definitions, we see immediately that the following condition is necessary and sufficient : for any point $a \in X$, for any local coordinate z_i of the first structure in a neighbourhood of a, and for any local coordinate w_α of the second structure in a neighbourhood of the same point a, w_α is expressible as a holomorphic function of z_i, and conversely z_i is expressible as a holomorphic function of w_α. This condition can also be expressed as follows : consider the open covering of X consisting of all the U_i and all the V_α with local coordinates z_i and w_α; then the required condition is that this system satisfies the coherence condition (1. 1); in other words, the system must define a complex manifold structure on X (which structure will be isomorphic both to that defined by the U_i and the z_i and to that defined by the V_α and the w_α). When two complex manifold structures on X satisfy the above condition we say that they are *equivalent*. A *complex manifold* is defined to be the datum of a Hausdorff topological space X and a class of analytic structures on X which are equivalent to each other.

Definition. Let X be a space with a complex manifold structure defined by the U_i and the z_i. Let U be an open set of X; the complex *structure* on U *induced* by that on X is defined to be the structure given by the open sets $U \cap U_i$ and the restriction of the functions z_i to $U \cap U_i$. In other words, if $a \in U$, then a local coordinate in a neighbourhood of a for the induced structure is merely a local coordinate in a neighbourhood of a for the given structure on X. Hence, any open set U of a complex manifold X is automatically provided with a complex manifold structure.

3. Examples of complex manifolds

Consider the plane **C** of the complex variable z. We take the covering formed by the single open set **C** and take z to be the local coordinate in **C**. This defines a complex manifold structure on **C** because the coherence condition (1. 1) is trivially satisfied. In accordance with the end of no. 2, any open set $D \subset \mathbf{C}$ is provided with a complex manifold structure; with this definition, the holomorphic functions on the complex manifold D are merely what we have always called holomorphic functions in D. As a second example of a complex manifold, we have the Riemann sphere which we have already discussed (chapter III, § 5, no. 1).

Consider now the quotient space $\mathbf{C}/\mathbf{Z}$ of the plane **C** by the additive subgroup **Z** of real points with integral coordinates. A point of $\mathbf{C}/\mathbf{Z}$

is then an equivalence class formed by points whose differences are integers. According to the general definitions of topology, $\mathbf{C}/\mathbf{Z}$ is provided with the *quotient topology* of the topology of $\mathbf{C}$: a set of $\mathbf{C}/\mathbf{Z}$ is *open* if its inverse image in $\mathbf{C}$ (for the canonical mapping $\mathbf{C} \to \mathbf{C}/\mathbf{Z}$) is an open set of $\mathbf{C}$. This is equivalent to saying that the open sets of $\mathbf{C}/\mathbf{Z}$ are the images of open sets of $\mathbf{C}$ under the canonical mapping $\mathbf{C} \to \mathbf{C}/\mathbf{Z}$.
It is very easy to show that the topology of $\mathbf{C}/\mathbf{Z}$ is Hausdorff. To define a complex manifold structure on $X = \mathbf{C}/\mathbf{Z}$, consider the open sets V of $\mathbf{C}$ which are so small that the restriction to V of the canonical mapping $p : \mathbf{C} \to X$ is injective (for example, the open sets V of diameter < 1). If z denotes the coordinate in the plane $\mathbf{C}$, consider the pair formed by the open set $U = p(V)$ of X and the function $z \circ p^{-1}$ defined in this open set; we shall show that these pairs define a complex manifold structure on X. It is sufficient to check the coherence condition. Let, therefore, V_1 and V_2 be two sufficiently small open sets of $\mathbf{C}$ such that their images $U_1 = p(V_1)$ and $U_2 = p(V_2)$ intersect; call the restriction of p to V_i (for $i = 1, 2$) p_i, and put

$$V'_i = (p_i)^{-1}(U_1 \cap U_2) \subset V_i \quad (i = 1, 2).$$

The coherence condition requires that the mapping $f_{12} = p_1^{-1} \circ p_2$ of V'_2 on V'_1 is holomorphic and that its derivative is $\neq 0$. However, this is indeed the case because, if $z \in V'_2$, then $f_{12}(z)$ and z have the same image in $\mathbf{C}/\mathbf{Z}$, so, in some neighbourhood of each point of V'_2, $f_{12}(z) - z$ is a fixed integer.

We obtain another example of a complex manifold by considering a discrete subgroup Ω of $\mathbf{C}$ having as base a system of two vectors e_1 and e_2 with a ratio which is not real, as in chapter v, § 2, no. 5. Let X be the quotient $\mathbf{C}/\Omega$ provided with the quotient topology of the topology of $\mathbf{C}$; the open sets of X are the images of open sets of $\mathbf{C}$ by the canonical mapping $p : \mathbf{C} \to \mathbf{C}/\Omega$. The topology of X is Hausdorff and its complex manifold structure is defined exactly as in the previous example. But, in this case, the *space* $X = \mathbf{C}/\Omega$ *is compact* : for, consider a closed parallelogram of periods, say P; P is a compact subset of $\mathbf{C}$, so its image under p is a compact subset of X; but this image is the whole of X, and X is therefore compact. We have thus an example of a compact complex manifold other than the one we already know, the Riemann sphere.

4. Principle of analytic continuation; maximum modulus principle

The principle of analytic continuation (chapter I, § 4, no. 3, corollary 2) extends to holomorphic functions on a complex manifold and, more generally, to holomorphic mappings of a complex manifold X into a complex manifold X'. To be precise, let D be a non-empty open set of a *connected* complex manifold X; if two holomorphic mapping f and g of

X into X′ coincide on D, then *they coincide everywhere in* X. To prove this, it is sufficient to prove the following :

PROPOSITION 4. 1. *Let f and g be two holomorphic mappings of a complex manifold* X *into a complex manifold* X′; *the set* U *of points of* X *in a neighbourhood of which f and g coincide is both open and closed.*

Proof. By definition, U is open, and it is therefore sufficient to show that U is closed. Let a be a point of X in the closure of U; since f and g are continuous and coincide in U, we have $f(a) = g(a)$. In some neighbourhood of a in X, take a local coordinate which is zero at the point a; take also a local coordinate w in a neighbourhood of $f(a)$ in X′. In some neighbourhood of a, the functions $w \circ f$ and $w \circ g$ can be expressed as holomorphic functions $\varphi(z)$ and $\psi(z)$ in a neighbourhood V of $z = 0$. By the classical principle of analytic continuation, the set E of $z \in V$ in a neighbourhood of which φ and ψ coincide is closed; since the point $z = 0$ is in the closure of E, we have $0 \in E$; hence, φ and ψ coincide in a neighbourhood of 0, and, consequently, f and g coincide in a neighbourhood of the point $a \in X$.

This completes the proof.

PROPOSITION 4. 2. *Let f be a holomorphic function on a connected complex manifold* X. *If* $|f|$ *has a relative maximum at a point* $a \in X$, *then the function f is constant* (maximum modulus principle).

Proof. Consider a local coordinate z in a neighbourhood of a; the function f is expressed in a neighbourhood of a as a holomorphic function of z. Since $|f|$ has a relative maximum at the point a, the function f is constant in some neighbourhood of a by the classical maximum modulus principle (cf.chapter III, § 2, theorem 1). We see then, by the usual argument, that the set of points of X where f takes the value $f(a)$ is both open and closed; since X is connected, this set is the whole of X.

This completes the proof.

COROLLARY. *If* X *is a compact, connected, complex manifold, any holomorphic function on* X *is constant.*

For, $|f|$ is a continuons function on the compact space, so it attains its upper bound; by proposition 4. 2, the function f is then constant on X.

Examples. The Riemann sphere $\mathbf{S}_2$ and the space $\mathbf{C}/\Omega$ (cf. no. 3) are compact, connected, complex manifolds. Thus, any holomorphic function on one of these spaces is constant. By noting that the mapping $f \to f \circ p$ establishes a bijective correspondence between the holomorphic functions on $\mathbf{C}/\Omega$ and the holomorphic functions on $\mathbf{C}$ which have the points of Ω as periods, we obtain a result which we established earlier by another method, namely : *any holomorphic function on* $\mathbf{C}$ *which is doubly periodic is constant* (cf. chapter III, § 5, no. 5, corollary to proposition 5. 1).

5. Meromorphic functions on complex manifolds

Definition. Let X be a complex manifold; a *meromorphic function* on X is defined to be a holomorphic mapping of X into the Riemann sphere $\mathbf{S}_2$; a meromorphic function is simply a continuous function which can take the value ∞ and which in a neighbourhood of each point $a \in$ X, can be expressed as a meromorphic function of a local coordinate in the neighbourhood of a.

Let Ω again be a discrete subgroup of $\mathbf{C}$ generated by two elements e_1 and e_2 whose ratio is not real. The canonical mapping $\mathbf{C} \to \mathbf{C}/\Omega$ obviously sets up a bijective correspondence between the meromorphic functions on the complex manifold $\mathbf{C}/\Omega$ and the meromorphic functions on $\mathbf{C}$ having Ω as a system of periods.

6. Ramification index of a holomorphic mapping

Let φ : X $\to$ Y be a *holomorphic* mapping of a complex manifold X into a complex manifold Y, and let a be a point of X. Let z be a local coordinate in X in a neighbourhood of a and let w be a local coordinate in Y in a neighbourhood of $b = \varphi(a)$. Since φ is holomorphic, $w(\varphi(x))$, for x near a, can be expressed as a holomorphic function $f(z)$ of the local coordinate z. Suppose, for simplicity, that z is zero at the point a, and w vanishes at the point b.

Let p be the order of multiplicity of the root 0 of the equation $f(z) = 0$. It is easy to see that this integer p is independant of the choice of local coordinate z in the neighbourhood of a and of the choice of local coordinate w in the neighbourhood of b; for, the changes of local coordinates are effected by holomorphic functions whose derivatives are $\neq 0$.

The integer p thus defined is called the *ramification index* of the mapping φ : X$\to$Y at the point $a \in$ X. By § 1 (nos. 1 and 2), if p is the ramification index there exists a local coordinate z in a neighbourhood of a and a coordinate w in a neighbourhood of b, such that the transformation φ expressed in terms of these local coordinates is $w = z^p$. Conversely, if this is the case, the ramification index at the point a is equal to p. We see that, in a neighbourhood of a, the function φ takes each value in Y sufficiently near to b and distinct from b exactly p times. In particular, *a necessary and sufficient condition for the restriction of* φ *to a sufficiently small neighbourhood of* a *to be a homeomorphism of this neighbourhood on its image* (in other words, that φ is *locally simple* in a neighbourhood of a) *is that the rami-*

fication index p is equal to 1; we say then that the mapping φ is *uuramified* at the point a.

PROPOSITION 6. 1. *Any simple holomorphic mapping of a complex manifold* X *on a complex manifold* Y *is an isomorphism.*

For, by the previous argument, the ramification index is necessarily equal to 1 at any point $a \in X$; if $b = \varphi(a)$, the inverse mapping φ^{-1} is obtained in a neighbourhood of b by expressing the local coordinate z in a neighbourhood of a as a holomorphic function of the local coordinate w in a neighbourhood of b. This proves the proposition.

Example. Consider the mapping $z \to e^{2\pi i z}$, which is a holomorphic mapping of the additive group $\mathbf{C}$ onto the multiplicative group $\mathbf{C}^*$ of complex numbers $\neq 0$. By passing to the quotient, it induces a holomorphic mapping φ of the analytic space $\mathbf{C}/\mathbf{Z}$ onto $\mathbf{C}^*$. It is clear that this mapping is holomorphic and simple. It follows that φ is an *isomorphism* of $\mathbf{C}/\mathbf{Z}$ onto $\mathbf{C}^*$. In fact, it is also an isomorphism of the topological groups $\mathbf{C}/\mathbf{Z}$ and $\mathbf{C}^*$ as we saw in chapter I, § 3.

7. FUNDAMENTAL THEOREM OF CONFORMAL REPRESENTATION

Here we shall state, without proof, a theorem which generalizes the fundamental theorem, stated and proved in § 3 for open sets of the plane $\mathbf{C}$, to the case of complex manifolds.

FUNDAMENTAL THEOREM. *Any simply connected complex manifold* X *is isomorphic to one of the following three manifolds :*

1° *the Riemann sphere* $\mathbf{S}_2$;

2° *the plane* $\mathbf{C}$;

3° *the unit disc* $|z| < 1$.

The proof of this theorem is too difficult to be given here. Note that only one of the three analytic manifolds above is compact, the first $\mathbf{S}_2$. We deduce the following corollary :

COROLLARY. *Any compact simply-connected, complex manifold is isomorphic to the Riemann sphere. Any non-compact, simply-connected, complex manifold is isomorphic to the plane* $\mathbf{C}$ *or the unit disc (these two cases are mutually exclusive).*

8. INTEGRATION OF DIFFERENTIAL FORMS AND THE RESIDUE THEOREM

Definition of a holomorphic differential form on a complex manifold X : such a form is defined by giving a holomorphic differential form

$$\omega_i = f_i(z_i)\,dz_i$$

in each open set U_i with local coordinate z_i, where f_i is a holomorphic function in the open set $A_i \subset \mathbf{C}$, which is the image of U_i under the local coordinate z_i. The forms ω_i are further restricted by the following coherence condition : if z_i and z_j are two local coordinates in a neighbourhood of the same point $a \in X$, the differential form ω_j is deduced from the differential form ω_i by the change of variable

$$z_i = f_{ij}(z_j), \tag{8. 1}$$

where transformation (8. 1) is that which expresses the local coordinate z_i as a function of the local coordinate z_j. In other words, we must have the relation

$$f_j(z_j) = f_i(f_{ij}(z_j))\,f'_{ij}(z_j). \tag{8. 2}$$

We shall indicate quickly, without proofs, how the theory of holomorphic differential forms in an open set of the plane $\mathbf{C}$ can be generalized to the case we have in mind of holomorphic differential forms on a complex manifold. Let ω be a holomorphic differential form on a complex manifold X; in a neighbourhood of each point of X, there exists a *primitive* of ω, that is, a holomorphic function g such that $dg = \omega$. Such a primitive is determined up to the addition of a constant. A *global* primitive of ω does not exist in general; however, if the space X is simply connected, any holomorphic differential form on X has a primitive. In the general case where X is not simply connected, the integral of ω along a closed path of X is not always zero; this integral has the same value for two *homotopic* closed paths (in the sense of chapter II, § 1, no. 6). The value of the integral along such a closed path is called a *period* of the integral $\int \omega$.

Let X be an analytic manifold; we have, in some neighbourhood of each point a of X, a concept of *orientation* because each local coordinate in a neighbourhood of a defines a homeomorphism of a neighbourhood of a onto an open set of the plane $\mathbf{C}$, which has its natural orientation, and two local coordinates in a neighbourhood of a indeed define the same orientation, since the change of local coordinates is expressed by a holomorphic transformation. From this, the idea of the oriented boundary of a compact set contained in X can easily be deduced; if Γ is the oriented boundary of a compact set, then the integral $\int_\Gamma \omega$ is zero for any holomorphic differential form ω.

We shall now define the idea of a *residue* of a holomorphic differential form. Let E be a closed discrete subset of a complex manifold X (so that E consists of *isolated* points) and let ω be a holomorphic differential form in the complement of E. Let a be a point of E and let z be a local coordinate in a neighbourhood of a which is zero at the point a. In some neighbourhood of a, the form ω can be written $f(z)\,dz$, where f is holomorphic in the neighbourhood of 0 except perhaps at $z = 0$. The Laurent expansion of $f(z)$ shows that, in a neighbourhood of a, the form ω can be written

$$\omega = \omega_1 + \left(\frac{c_1}{z} + \frac{c_2}{z^2} + \cdots\right)dz, \tag{8.3}$$

where ω_1 is a holomorphic differential form in a neighbourhood of a (a included). Let γ be a closed path situated in a small neighbourhood of a which does not pass through a and whose index with respect to a is equal to 1 (the index of a closed path is defined by considering the image of a neighbourhood of a under a local coordinate). The classical residue theorem shows that

$$\int_\gamma \omega = 2\pi i\, c_1. \tag{8.4}$$

Hence, the coefficient c_1 which occurs on the right hand side of (8. 3) does not depend on the choice of local coordinate z, which is zero at the point a. We call it the *residue* of the differential form ω at the point a. Starting from this definition and reasoning exactly as in chapter III (§ 5, no. 2) we can prove the following :

THEOREM OF RESIDUES. *If the oriented boundary* Γ *of a compact set* K *does not contain any of the points of the discrete closed set* E *in the complement of which the differential form* ω *is holomorphic, the integral* $\int_\Gamma \omega$ *is equal to* $2\pi i$ *times the sum of the residues of* ω *at the points of* E *situated in* K.

5. Riemann Surfaces

1. DEFINITIONS

Definition. Let Y be a complex manifold; a *Riemann surface spread over* Y (or, simply, *a Riemann surface over* Y) is defined to be a *connected* complex manifold X and a *non-constant* holomorphic mapping $\varphi : X \to Y$. We usually consider the case where Y is the plane $\mathbf{C}$ of the complex variable, or the Riemann sphere $\mathbf{S}_2$, in other words, Riemann surfaces spread over the plane, or the sphere.

We have seen in § 4, no. 6 the effect of the mapping φ in a neighbourhood of a general point $a \in X$: if the ramification index of φ is equal to 1 at the point a, φ defines a homeomorphism of a neighbourhood of a onto a neighbourhood of $\varphi(a)$; if the ramification index of φ at the point a is an integer $p > 1$, the image under φ of a small neighbourhood of a covers a neighbourhood of $\varphi(a)$ p times. The ramifications of φ (points where the ramification index is > 1) are *isolated* points of X. The mapping φ is always an *open* mapping and the inverse image of a point of Y is a *discrete* subset of X.

It should be understood that the mapping φ is not necessarily injective even if there are no ramifications; moreover, when there are ramifications, the image under φ of the (discrete) set of ramifications is not necessarily a discrete subset of Y; it can also happen that an infinity of distinct ramifications of X have the same image in Y.

Definition. An *unramified Riemann surface* over Y is a Riemann surface (X, φ) where the mapping φ is unramified, that is to say, where φ has ramification index 1 at each point of X.

To define an unramified Riemann surface over Y, it is sufficient to take a connected, *Hausdorff*, topological space X and a continuous mapping φ of X into Y which is *locally a homeomorphism* (this means that any point of X has an open neighbourhood V such that the restriction of φ to V is a homeomorphism of V onto its image $\varphi(V)$). For, the mapping φ then defines, in a neighbourhood of each point of X, a *local coordinate*, and so X is provided with a complex manifold structure and it is clear that the mapping φ is a holomorphic mapping of X in Y. A particular case of an unramified Riemann surface over Y is that of a *covering space* of Y :

Definition. A *covering space* of Y is an unramified Riemann surface (X, φ) which satisfies the following condition :

For any point $b \in Y$, there exists an open neighbourhood V of b in Y such that the inverse image $\varphi^{-1}(V)$ is composed of disjoint open sets U_i of X, each of which is mapped homeomorphically onto V by the mapping φ.

Example. Let $Y = \mathbf{C}^*$, the complement of 0 in the plane $\mathbf{C}$. Let $X = \mathbf{C}$ and let $z = e^t$ be the mapping of X into Y. This mapping makes X into a covering space of Y. For, let b be a complex number $\neq 0$, and take an open disc of centre b and radius $< |b|$ for V. Each branch of $\log z$ in V is a function which defines a homeomorphism of V onto an open set of the plane $\mathbf{C}$. These open sets U_i are mutually disjoint, their union is $\varphi^{-1}(V)$, and the restriction of φ to each U_i is a homeomorphism of U_i on V.
In the above example, the space $Y = \mathbf{C}^*$ is not simply connected, but its covering space $\mathbf{C}$ is simply connected. This is therefore an example of a connected but not simply connected manifold having a simply connected covering space. We state the following theorem without proof :

THEOREM. *Any connected open set of the plane $\mathbf{C}$ (or, more generally, any connected complex manifold Y) has a simply connected covering space.*

2. Holomorphic functions and holomorphic differential forms on a Riemann surface

Definition. Given a Riemann surface (X, φ) over Y, a *holomorphic* (resp. *meromorphic*) *function* on this Riemann surface is simply defined to be a holomorphic (resp. meromorphic) function on the analytic manifold X. A holomorphic differential form on a Riemann surface is defined similarly.

For example, consider the Riemann surface mentioned at the end of no. 1 : $X = \mathbf{C}$ (with complex variable t), $Y = \mathbf{C}^*$ (with complex variable $z \neq 0$), and the mapping φ is given by $z = e^t$. Since the mapping φ is unramified, we can take the function $e^t = z$ as a local coordinate in a neighbourhood of each point of X; then any holomorphic function f on the Riemann surface can be expressed *locally* as a holomorphic function of z. But, since different points of X can be mapped by φ onto the same point of Y, f is not in general a global (single-valued) holomorphic function of the variable $z \neq 0$. In particular, t is a holomorphic function on X; in a neighbourhood of each point of X, t is one of the branches of log z, but log z is not a single-valued function of z on $\mathbf{C}^*$. We can say that we have *made* log *z single-valued* by considering the covering space $\varphi : X \to Y$: instead of considering it as a function on $\mathbf{C}^*$, we consider it as a function on the simply connected covering space (X, φ) of $\mathbf{C}^*$.

We shall now study another example of how a many-valued function can be 'made single-valued' by the introduction of a suitable Riemann surface (we shall not deal with the general case). Consider the many-valued function

$$y = (1 - x^3)^{1/3}$$

in the plane $\mathbf{C}$. At each point x other than $1, j$ and j^2 (the cube roots of 1), y has three distinct values; if x takes one of the values $1, j$, or j^2, the three values of y coincide at the value zero. We intend to define a complex manifold X and a holomorphic mapping $\varphi : X \to \mathbf{C}$. Consider the product $\mathbf{C} \times \mathbf{C}$ consisting of pairs (x, y) of complex numbers and the subset X consisting of pairs such that

$$x^3 + y^3 = 1. \tag{2. 1}$$

We provide X with the topology induced from the product $\mathbf{C} \times \mathbf{C}$; thus X is a Hausdorff topological space. We consider the two functions on X, denoted by x and y, which are, in fact, the first and second coordinates of the point (x, y). To define a complex manifold structure on X, we choose a local coordinate in a neighbourhood of each point of X. Take first a point $(x_0, y_0) \in X$ such that $y_0 \neq 0$ (so x_0 is neither 1, j, nor j^2); then x is taken as the local coordinate. This choice is valid because the function x defines a homeomorphism of a neighbourhood of the point (x_0, y_0) (in X) onto a neighbourhood of x_0 (in $\mathbf{C}$) : the inverse homeomorphism takes a point x (near to x_0) to the pair (x, y), where $y = (1 - x^3)^{1/3}$ the branch being that which is equal to y_0 at $x = x_0$. Now let (x_0, y_0) be a point of X such that $y_0 = 0$; then $x_0 \neq 0$ (in fact, x_0 is one of the numbers $1, j, j^2$) and we can take y as a local coordinate; this choice is valid for the same reason as in the other case (but with the roles of x and y interchanged). We still need to check that the local coordinates thus defined satisfy the coherence condition ((1. 1) of § 4). In other words, we must verify that in a neighbourhood of a point $(x_0, y_0) \in X$ such that $x_0 \neq 0$ and $y_0 \neq 0$, equation (2. 1) defines the local coordinate y as a holomorphic function of the local coordinate x, and *vice-versa*. However, this is indeed the case because $(1 - x^3)^{1/3}$ has a holomorphic branch which is equal to y_0 for $x = x_0$; similarly, $(1 - y^3)^{1/3}$ has a holomorphic branch which is equal to x_0 for $y = y_0$.

Hence we have provided the topological space X with a complex manifold structure. Each of the functions x and y is a holomorphic function on X for this structure. Let us verify this for x : it is obvious for a point (x_0, y_0) such that $y_0 \neq 0$ because x is a local coordinate; at a point (x_0, y_0) such that $y_0 = 0$, y is a local coordinate, and $x = (1 - y^3)^{1/3}$ is a holomorphic function of y.

Let the mapping $\varphi : X \to \mathbf{C}$ be the function x. Thus, (X, φ) is a Riemann surface over $\mathbf{C}$ and it is the one that we have been seeking. On this Riemann surface, $y = (1 - x^3)^{1/3}$ is indeed a (single-valued) holomorphic function. We note that the Riemann surface X has three points over each point $x \in \mathbf{C}$, the three points (x, y) such that $y = (1 - x^3)^{1/3}$. We say that the Riemann surface has three 'sheets'. But these three points coincide if x is one of the points $1, j, j^2$ of $\mathbf{C}$.

We define a particular *holomorphic differential form* ω on the above complex manifold X as follows: in a neighbourhood of a point (x_0, y_0) such that $y_0 \neq 0$, we put

$$\omega = \frac{dx}{y},$$

and, in a neighbourhood of a point (x_0, y_0) such that $y_0 = 0$ (so $x_0 \neq 0$), we put

$$\omega = -\frac{y\,dy}{x^2},$$

(If $x_0 \neq 0$ and $y_0 \neq 0$, the relation

$$x^2\,dx + y^2\,dy = 0,$$

which is a consequence of (2. 1), implies the equality of the differential forms $\frac{dx}{y} = -\frac{y dy}{x^2}$). We can say that ω is simply the differential form

$$\frac{dx}{(1 - x^3)^{1/3}}$$

on $\mathbf{C}$, made holomorphic by introducing the Riemann surface (X, φ) over $\mathbf{C}$.

Example. Show that the closed path γ of the plane $\mathbf{C}$, illustrated in figure 12, is actually the image under φ of a closed path on the Riemann surface X; in fact, there are *three* closed paths on X whose image is γ. By integra-

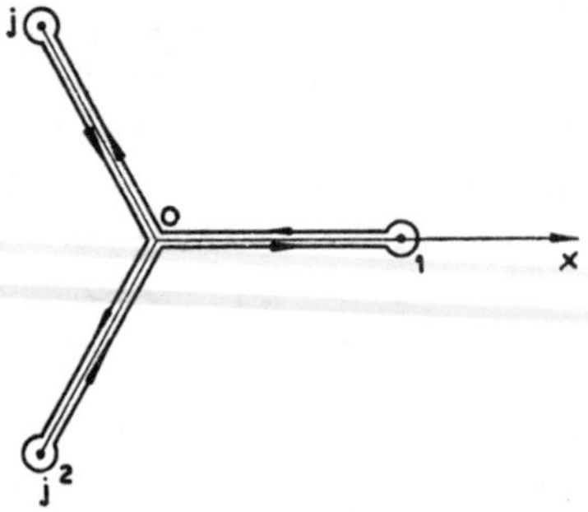

Fig. 12

ting the above differential form ω round one of them, show that the real integral $\int_0^1 \frac{dx}{(1 - x^3)^{1/3}}$ is equal to $\frac{2\pi}{3\sqrt{3}}$.

Let us reconsider equation (2. 1). It will lead us to define a Riemann surface, not over $\mathbf{C}$, but over the Riemann sphere $\mathbf{S}_2$. Consider then the *complex projective plane* $P_2(\mathbf{C})$, which is the quotient of

$$\mathbf{C} \times \mathbf{C} \times \mathbf{C} - \{0, 0, 0\}$$

by the following equivalence relation :

$(x, y, z) \sim (x', y', z')$ if x, y, z are proportional to x', y', z' (we say that (x, y, z) is a system of homogeneous coordinates of the point of $P_2(\mathbf{C})$ which it defines, that is to say, of its equivalence class). The points of $P_2(\mathbf{C})$ whose homogeneous coordinate x, y, z satisfy the equation

$$x^3 + y^3 = z^3 \tag{2. 2}$$

form a Hausdorff topological space X′. The space X above can be identified with a subspace of X′ by associating with each point $(x, y) \in X$ the point with homogeneous coordinates $(x, y, 1)$. The space X′ is composed of the space X and three points at 'infinity' : these points have homogeneous coordinates $(1, -1, 0)$, $(j, -1, 0)$ and $(j^2, -1, 0)$. The reader should define a complex manifold structure on X′, which extends that of X (it is sufficient to define a local coordinate in a neighbourhood of each of the three points at infinity of X′), and a holomorphic mapping $\varphi' : X' \to \mathbf{S}_2$, which extends φ. It can be shown that X′ is a compact complex manifold and that y/z is a meromorphic function on X′, whose poles are the three points at infinity.

3. Riemann surface associated with an 'elliptic curve'

Consider an algebraic relation between two complex variables x and y, of the form

$$y^2 = 4x^3 - 20a_2x - 28a_4 \tag{3. 1}$$

(we supposely use the same notation as in chapter v, § 2, n° 5). We assume that a_2 and a_4 are chosen so that the polynomial on the right hand side of (3. 1) has three distinct roots. Write this polynomial $P(x)$; then $P'(x) \neq 0$ for any value of x such that $P(x) = 0$, P' denoting the derivative of P. We shall associate a Riemann surface (X, φ) over $\mathbf{C}$ with the 'elliptic curve' (3. 1). The topological space X is a subspace of $\mathbf{C} \times \mathbf{C}$ defined by the pairs (x, y) which satisfy (3. 1). The complex manifold structure the space X is defined as follows: at a point $(x_0, y_0) \in X$ such that $y_0 \neq 0$, we take x as 'local coordinate'; at a point (x_0, y_0) such that $y_0 = 0$, we have $P'(x_0) \neq 0$, thus by the implicit function theorem relation (3. 1) is equivalent to a relation of the form $x = f(y)$ in a neighbourhood of $(x_0, 0)$, where f is holomorphic in some neighbourhood of 0 and $f(0) = x_0$. We take y as a local coordinate in a neighbourhood of such a point $(x_0, 0)$.

The mapping $\varphi: X \to \mathbf{C}$ which maps (x, y) onto $x \in \mathbf{C}$ is obviously holomorphic. Thus (X, φ) is a Riemann surface over $\mathbf{C}$; it has two 'sheets' because a value of x corresponds to two distinct values of y in general (they are distinct if $P(x) \neq 0$). Moreover, the function $X \to \mathbf{C}$ which maps the pair (x, y) onto y is also holomorphic on X; we shall write it simply y.

The differential form ω, defined by $\omega = dx/y$ in a neighbourhood of points $(x_0, y_0) \in X$ such that $y_0 \neq 0$, and by $\omega = \dfrac{dy}{6x^2 - 10a_2}$ in a neighbourhood of points $(x_0, 0) \in X$, is a holomorphic differential form on X. Since it is a closed form, it has a primitive in a neighbourhood of each point of X; globally, this primitive is a *many-valued* function z which is holomorphic in a neighbourhood of each point of X. The equation $dz = \omega$ shows that

$$dx = y\, dz. \tag{3. 2}$$

Each branch of the function z in a neighbourhood of each point $(x_0, y_0) \in X$ is a local coordinate: for, if $y_0 \neq 0$, x is a local coordinate and $dz = \dfrac{1}{y} dx$; if $y_0 = 0$, y is a local coordinate and $dz = \dfrac{dy}{6x^2 - 10a_2}$, the denominator not being zero.

We shall now complete the Riemann surface (X, φ') to obtain a Riemann surface (X', φ') over the Riemann sphere $\mathbf{S}_2$. To do this, we let (x, y, t) be *homogeneous coordinates* of a point of the complex projective plane $P_2(\mathbf{C})$ (cf. n° 2) and we consider the set X' of points of $P_2(\mathbf{C})$ whose homogeneous coordinates satisfy

$$y^2t = 4x^3 - 20a_2xt^2 - 28a_4t^3. \tag{3. 3}$$

The topological space X' is Hausdorff; we identify X with a subspace of X' by associating the point $(x, y) \in X$ with the point of X' whose homogeneous coordinates are $(x, y, 1)$; the complement $X' - X$ consists of a single point at infinity, the point with homogeneous coordinates $(0, 1, 0)$. In a neighbourhood of this point, denoted by ∞, we can take $x/y = x'$ as local coordinate because x' defines a homeomorphism of a neighbourhood of the point ∞ onto a neighbourhood of 0 in $\mathbf{C}$ (for, put $t/y = t'$; equation (3. 3) is equivalent to

$$t' = 4x'^3 - 20a_2x't'^2 - 28a_4t'^3;$$

in some neighbourhood of $x' = 0$, $t' = 0$, the implicit function theorem gives t' as a holomorphic function of x':

$$(3.4) \qquad t' = 4x'^3 - 320a^2x'^7 + \cdots).$$

A complex manifold structure is defined on X′ when we have chosen x' as the local coordinate at ∞ (the reader should check the coherence conditions). Finally, the mapping φ' is defined to be equal to φ on X and to take the point ∞ of X′ onto the point at infinity of $\mathbf{S}_2$.

The holomorphic differential form ω, which we defined above on X, extends to a holomorphic differential form on X′; in a neighbourhood of the point ∞, we use the local coordinate x' and the holomorphic function t' of x' defined by (3.4) and we put

$$\omega = t'd(x'/t') = dx' - x'\frac{dt'}{t'} = dx' - \frac{12\,x'^2 + \cdots}{4\,x'^2 + \cdots}\,dx' = -\,2\,dx'(1 + g(x')),$$

where g is a holomorphic function in a neighbourhood of $x' = 0$ and is zero for $x' = 0$. The form ω thus defined on the (compact) space X′ has locally a primitive, which is a many-valued function on X′ and serves as a local coordinate at each point of X′.
Suppose now that the constants a_2 and a_4 are obtained from a discrete group Ω by the relations (5.5) of chapter v, § 2. Then, proposition 5.2 of the same paragraph shows that the meromorphic transformation

$$(3.5) \qquad x = \wp(z), \qquad y = \wp'(z)$$

defines an *isomorphism* of the complex manifold $\mathbf{C}/\Omega$ onto the complex manifold X′. The inverse isomorphism defines z as a holomorphic many-valued function on X′, whose (local) branches differ in value by a constant belonging to Ω. We have $dx = y\,dz$ because of (3.5), and, consequently, dz (which is a well-defined differential form on X′) is simply the form ω defined above (which justifies the notation z).

Let us now abandon the hypothesis that a_2 and a_4 arise from a discrete subgroup Ω by the formulae (5.5) of chapter v, § 2. The many valued function z is still defined on X′ by the condition that $dz = \omega$; a more searching analysis of the topology of the space X′ would reveal that the different branches of z are obtained from one another by adding constants which form a discrete subgroup Ω of the additive group $\mathbf{C}$, and that Ω is generated by two elements e_1 and e_2 which are linearly independent over the real field $\mathbf{R}$. We can fix the many-valued function z by imposing the condition that it is zero (mod. Ω) at the point ∞ of X′. We then introduce the elliptic curve

$$y^2 = 4x^3 - 20b_2x - 28b_4,$$

where the constants b_2 and b_4 are given by

$$b_2 = 3\sum_{\omega\in\Omega,\ \omega\neq 0}\frac{1}{\omega^4}, \qquad b_4 = 5\sum_{\omega\in\Omega,\ \omega\neq 0}\frac{1}{\omega^6}.$$

Let (X″, φ'') be the corresponding Riemann surface over $\mathbf{S}_2$. The many-valued function z defines a holomorphic mapping of X′ into $\mathbf{C}/\Omega$, which, as we have just seen, is isomorphic with X″; we have therefore a holomorphic mapping f: X′→X″. It can be shown (though we shall not do so here) that f is an isomorphism; hence, f takes each class of values mod. Ω once and once only on X′. Thus the (non-homogeneous) coordinates x and y of a point of X′ are meromorphic functions of z with Ω as group of periods, and, since it can easily be seen that x is function of z with a double pole at $z = 0$ and with $1/z^2$ as its principal part (and having no poles other than those at points of Ω), it follows that $x = \wp(z)$, where $\wp$ is the

Weierstrass function attached to the group Ω, and $y = \wp'(z)$. Hence, the isomorphism $f : X' \to X''$ is merely the identity mapping, and we have $b_2 = a_2$, $b_4 = a_4$. Finally, we conclude that *any pair of constants* a_2 *and* a_4, *such that the polynomial* $P(x)$ *on the right hand side of* (3. 1) *has three distinct roots, defines a discrete group* Ω *such that* a_2 *and* a_4 *satisfy the relations* (5. 5) *of chapter* v, § 2; *moreover, the algebraic elliptic curve* (3. 1), *including its point at infinity, has a parametric representation given by formulae* (3. 5).

4. Some ideas on analytic continuation

We shall confine our attention to analytic continuation in the complex plane $\mathbf{C}$. We formulate the problem as follows :

Problem. Suppose we are given a non-empty open set U of the plane $\mathbf{C}$ (U will regarded as a Riemann surface over $\mathbf{C}$, the mapping $i : U \to \mathbf{C}$ being the inclusion mapping), and suppose that f is a given holomorphic function in U. We seek an unramified Riemann surface (X, φ) over $\mathbf{C}$ and an isomorphism j of U onto an open set of X with the following conditions :

(i) $\varphi \circ j = i$ (which enables us to identify U with a 'sub-Riemann surface' of X);

(ii) the function f extends to a holomorphic function g in X ('extends' means that $g \circ j = f$ in U);

(iii) the Riemann surface (X, φ) is the 'greatest possible' of those satisfying (i) and (ii). This means that, if (X', φ') is an unramified Riemann surface over $\mathbf{C}$ and j' is an isomorphism of U onto an open set of X' satisfying conditions similar to (i) and (ii), then *there exists a unique holomorphic mapping*

$$h : X' \to X$$

such that

$$h \circ j' = j, \qquad \varphi \circ h = \varphi'. \tag{4. 1}$$

Before proceeding further, we note that the holomorphic function g 'extending' f in condition (ii) is *unique* because of the 'principle of analytic continuation' (cf. § 4, no. 4). Moreover, if g' is the unique holomorphic function in the X' of condition (iii) such that $g' \circ j' = f$, then,

$$g \circ h = g'; \tag{4. 2}$$

for, $g \circ h$ is holomorphic in X', and we certainly have $(g \circ h) \circ j' = f$ in U because $h \circ j' = j$ by (4. 1) and $g \circ j = f$ by hypothesis.

We express property (iii) briefly saying that the triple (X, φ, j) is *universal* (with respect to triples satisfying (i) and (ii)).

The fundamental theorem of analytic continuation is this : given a connected open set $U \subset \mathbf{C}$ and a holomorphic function f in U, *the above problem has a solution which is unique.* By uniqueness here, we mean ' up to isomorphism '. Our first step is to prove this uniqueness and thus to clarify what we mean by ' up to isomorphism '.

Proof of uniqueness. Suppose that there are two solutions (X, φ, j) and (X_1, φ_1, j_1) to the problem. Since (X, φ, j) has the universal property, there exists a unique holomorphic mapping $h : X_1 \to X$ such that

$$(4.\ 3) \qquad h \circ j_1 = j, \qquad \varphi \circ h = \varphi_1.$$

For the same reason, there exists a unique holomorphic mapping $h_1 : X \to X_1$ such that

$$h_1 \circ j = j_1, \qquad \varphi_1 \circ h_1 = \varphi.$$

Consider the mapping $h \circ h_1$ of X into itself; it is a mapping k such that

$$k \circ j = j, \qquad \varphi \circ k = \varphi;$$

however, by the universal property, there is *only one* holomorphic mapping with these properties, and, since the identity mapping of X has these properties, it follows that $k = h \circ h_1$ is the identity mapping of X_1. For the same reason, $h_1 \circ h$ is the identity mapping of X_1. This implies that h_1 and h are isomorphisms and are inverse to one another.

Hence, any two solutions of the problem can be derived from each other by an isomorphism h satisfying (4. 3); it is in this sense that we say that the solution to the problem (if it exists) is *unique up to isomorphism.*

We have still to prove the *existence* of a solution and this is more tricky. The reader may wish to omit this proof in his first reading.

Let Z be the set of pairs (z_0, S) consisting of a point $z_0 \in \mathbf{C}$ and a power series S (in one variable) whose radius of convergence is non-zero. We define a topology on Z : to each pair (V, F) consisting of an open set $V \subset \mathbf{C}$ and a holomorphic function F in V, we associate the set W(V, F) of pairs (z_0, S) where $z_0 \in V$ and S is the power series $\sum_{n \geqslant 0} a_n X^n$ such that $\sum_{n \geqslant 0} a_n (z - z_0)^n$ is the power series expansion of F in a neighbourhood of z_0; we define the topology on Z by stipulating that the sets W(V, F) form a *base of open sets* for this topology, that is to say, any open set of Z is a union of sets of the form W(V, F), and, conversely, any union of sets of the form W(V, F) is open; thus, the topology on Z is well defined. This topology is *Hausdorff* : that two distinct points (z_0, S) and (z_0', S) have disjoint open neighbourhoods is obvious if $z_0 \neq z_0'$; and, if $z_0 = z_0'$ and $S \neq S'$, the set of points (sufficiently near to z_0) in a neighbourhood of which the holomorphic functions, defined by the distinct power series S and S′, coincide is *empty* because of the principle of analytic continuation.

Let $p : Z \to \mathbf{C}$ be the mapping which maps each pair (z_0, S) onto $z_0 \in \mathbf{C}$; it is *locally a homeomorphism* (i. e. any point of Z has a neighbourhood which is mapped homeomorphically by the restriction of p onto a neighbourhood of the image of the point) : this follows directly from the definition of the topology of Z. By using p as *a local coordinate* in a neighbourhood of each point of the space Z, we define a *complex manifold structure* on Z. The mapping p is a holomorphic mapping of this structure, so (Z, p) would be a Riemann surface over **C** if Z were connected (we shall see that this is not the case).

We define a function G on the space Z as follows : the value of G at a point $(z_0, S) \in Z$ is defined to be the constant term of the power series $S = \sum_{n \geqslant 0} a_n X^n$; this is also the value, at the point z_0, of the holomorphic function $\sum_{n \geqslant 0} a_n (z - z_0)^n$ defined in a neighbourhood of z_0. The definition of the topology of Z shows, clearly, that the function G thus defined on the space Z is holomorphic; in fact, if it is expressed as a function of the local coordinate Z in a neigbourhood of (z_0, S), then we find, that the function G has the power series expansion $\sum_{n \geqslant 0} a_n (z - z_0)^n$, where $\sum_{n \geqslant 0} a_n X^n$ is precisely the series S.

So far, we have not used the given non-empty, connected, open set U, or the holomorphic function f in U. We shall introduce them now. Consider W(U, f) : it is a (non-empty, connected) open set of the complex manifold Z by definition, and the restriction of the mapping $p : Z \to \mathbf{C}$ to this open set W(U, f) is an isomorphism of W(U, f) onto the open set $U \subset \mathbf{C}$. Let j be the inverse isomorphism. The composed mapping $G \circ j$ is simply f. Let X be the connected component of Z which contains the open set $j(U)$, let φ be the restriction of p to X, and let g be the restriction of G to X.

Since p is locally a homeomorphism, so is φ; thus (X, φ) is indeed an *unramified* Riemann surface over **C**. To show that (X, φ) and the isomorphism j satisfy the conditions of the problem, it remains to be checked that they satisfy conditions (i), (ii) and (iii). Condition (i) follows trivially from the definition of j. Condition (ii) is true because g is the restriction of G to X and because $G \circ j = f$ as we have seen. To prove (iii), let (X′, φ') and j' be as in (iii), with a holomorphic function g' in X′ such that $g' \circ j' = f$ in U; we define a mapping k of X′ into Z taking a point $x'_0 \in X'$ onto the pair $(\varphi'(x'_0), S)$, where S denotes the power series $\sum_{n \geqslant 0} a_n X^n$ such that $\sum_{n \geqslant 0} a_n (z - z_0)^n$ is the expansion in a neighbourhood of $z_0 = \varphi'(x'_0)$ of the holomorphic function $\lambda(z)$ defined by' $\lambda(\varphi'(x')) = g'(x')$ in a neighbourhood of x'_0 (we use here the condition that the Riemann surface (x', φ') is unramified and that $\varphi'(x')$ is therefore a local coordinate in X′ in a neighbourhood of x'_0). The mapping k, which we have just defined, is holomorphic in X′. Since the space X′ is connected (by the definition of a Riemann surface), its image by k is connected; and, as this image obviously contains the open set $j(U)$, it is contained in the connected component X of Z. Therefore, k induces a holomorphic mapping h of X′ into X and it is easily verified that h satisfies conditions (4. 1). To complete the proof, it remains to be proved that any holomorphic mapping $h : X' \to X$ satisfying (4. 1) coincides with the one defined above; they must coincide in the non-empty open set $j'(U)$ and, consequently, they coincide on the whole of the connected space X′ because of the principle of analytic continuation (§ 4, no. 4).

Comments. We have thus obtained a ' largest ' unramified Riemann surface over **C**, containing the given open set U and to which the function f can be extended holomorphically. The most simple idea would have been to seek a ' largest ' connected open set V containing U to which f can be extended holomorphically. However, this problem *has no solution* in general, and it is because of this that unramified Riemann surfaces, rather than, simply, open sets of **C**, must be considered in this context. Here is an example showing that there is no largest open set V containing U and allowing an extension of f: we take as U an open disc of **C** which does not contain o, and we take a branch of log z in U as the function f; let U′ be the symmetric image of U with respect to o; then, it is easy to construct two simply connected open sets V_1 and V_2 each containing both U and U′ (see figure 13) such that, if the branch of log z in U is extended

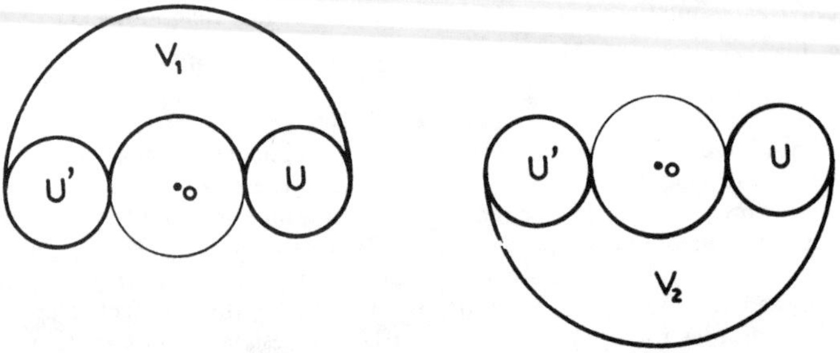

Fig. 13

in turn to V_1 and V_2, then the two extensions give different branches when restricted to U′. There is therefore no open set V containing V_1 and V_2 to which the branch of log z in U can be extended and therefore no largest open set containing U to which the extension is possible.

Let f again be a holomorphic function in an open set $U \subset \mathbf{C}$, and let $z_0 \in U$. Consider a path $\gamma : I \to \mathbf{C}$ with initial point z_0 and end point z_1 (I denotes the segment [o, 1]); we do not assume that the image of γ is contained in U. Moreover, let the triple (X, φ, g) satisfy conditions (i), (ii) and (iii). If there is a continuous mapping $h : \to X$ such that $\varphi \circ h = \gamma$ and $h(0) = j(z_0)$, then this h is unique (the proof is easy); it is called a *lifting* of the path γ in the Riemann surface (X, φ). In a neighbourhood of the point $h(1) \in X$, the function g is expressible as a holomorphic function of $z = \varphi(x)$ in some neighbourhood of the point z_1; we say that this holomorphic function of z has been obtained by *analytic continuation of the holomorphic function f along the path* γ (with initial point z_0 and end point z_1).

The above theory of analytic continuation has many generalizations. For example, instead of extending to an unramified Riemann surface over **C**, we can extend to an unramified Riemann surface *over the Riemann sphere*; the arguments are similar. It is also possible to consider Riemann surfaces without the restriction that they shall be unramified, with conditions similar to (i), (ii) and (iii); it can be shown that this problem also, has a solution which is unique 'up to isomorphism'.

Exercises

1. Let D be a simply connected open set of the plane **C**, different from **C**, and let $a \in \mathrm{D}$. Given a complex number b such that $|b| < 1$ and a real number α, show that there exists a *unique* holomorphic function $f(z)$ which defines an isomorphism of D on the unit disc and satisfies

$$\text{(i)} \quad f(a) = b, \qquad \text{(ii)} \quad \arg f'(a) = \alpha.$$

2. Let D be a connected open set whose frontier is the union of two non-intersecting circles C_1, C_2, such that C_1 is inside C_2. Show that there is a homographic transformation which maps D onto an annulus $r < |z| < 1$ (where r is a suitable number > 0 and < 1).

3. Let $f(z)$ be a simple holomorphic function in the unit disc B given by $|z| < 1$, and let $\mathrm{D} = f(\mathrm{B})$ be the image of B under f. Similarly, let D_r be the image under f of the open disc B_r given by $|z| < r$ for $0 < r < 1$.

(i) Show that, if h is an automorphism of D leaving the point $f(0)$ fixed, then

$$h(\mathrm{D}_r) \subset \mathrm{D}_r \qquad \text{for} \qquad 0 < r < 1.$$

(Apply Schwarz' lemma to the function $f^{-1}(h(f(z)))$.)

(ii) Show that, if D is starred with respect to the point $f(0)$ (see chapter II, § 1, no. 7 for the definition), then D_r is also starred with respect to the point $f(0)$ for $0 < r < 1$. (Reduce the problem to the case when $f(0) = 0$ and consider the function $f^{-1}(\lambda f(z))$ for $0 < \lambda < 1$.)

(iii) Suppose now that D is *convex*. Show that D_r is also convex for $0 < r < 1$. If E is an open disc such that $\overline{\mathrm{E}} \subset \mathrm{B}$, is the image $f(\mathrm{E})$ convex? (Given $0 < \lambda < 1$ and $z_1, z_2 \in \mathrm{B}$, consider the function

$$g_\lambda(z) = (1 - \lambda) f(z z_1/z_2) + \lambda f(z),$$

where $|z_1| \leq |z_2| \neq 0$. For the second part, show that there is an automorphism φ of B such that $\varphi(\mathrm{E}) = \mathrm{B}_r$, where $0 < r < 1$, and consider the function $f \circ \varphi^{-1}$.)

4. Let $w = f(z)$ be a simple holomorphic function in the unit disc $|z| < 1$ and let Γ be the image of the circle $|z| = r$, $0 < r < 1$, under f. Show that the radius of curvature ρ of Γ at a point $f(a)$, with $|a| = r$, is given by the following formula:

$$\frac{1}{\rho} = \frac{\operatorname{Re}(af''(a)/f'(a)) + 1}{|af'(a)|}.$$

(Note that, if $f(re^{i\theta}) = u(\theta) + iv(\theta)$, then

$$\frac{u'v'' - u''v'}{u'^2 + v'^2} = \operatorname{Im}\left(\frac{u'' + iv''}{u' + iv'}\right),$$

where u', u'', etc., denote derivatives with respect to θ.)

5. Let a be a complex number, r a real number and z_1, z_2 two corresponding points of the inversion with respect to the circle C of centre a and radius r. Let S be a homographic transformation whose pole does not lie on the circle C; show that $S(z_1)$ and $S(z_2)$ are corresponding points of some inversion, and determine its centre and radius. If the pole of S lies on C, show that $S(z_1)$ and $S(z_2)$ are corresponding points of the reflection in the line S(C).

6. Let C (resp. Γ) be a circle in the plane of the complex variable z (resp. w) defined by $|z - a| = r$ (resp. $|w - \alpha| = \rho$). Let D (resp. Δ) be a connected open set of the z-(resp. w-)plane satisfying the following conditions :

(i) $C_0 = D \cap C$ (resp. $\Gamma_0 = \Delta \cap \Gamma$) is a non-empty (open) arc of C (resp. Γ);

(ii) $D_+ = J_C(D_-)$ (resp. $\Delta_+ = J_\Gamma(\Delta_-)$),
where J_C (respc. J_Γ) denotes the inversion with respect to the circle C (resp. Γ) and $D_\pm$ (resp. $\Delta_\pm$) denotes the set of the $z \in D$ (resp. $w \in \Gamma$) such that $|z - a| \gtrless r$ (resp. $|w - \alpha| \gtrless \rho$).

Let there be also a function f defined and continuous in $D_+ \cup C_0$ with values in $\Delta_+ \cup \Gamma_0$ such that :

(iii) f is holomorphic in D_+ and maps D_+ into Δ_+;

(iv) f maps C_0 into Γ_0.

Show that, with these hypotheses, f can be extended uniquely to a holomorphic function g in D which maps D_- into Δ_-. (Reduce it to the case when C_0 (resp. Γ_0) is contained in the real axis and use Schwarz principle of symmetry, chapter II, § 2, no. 9).

We now replace hypotheses (iii) and (iv) by the more restrictive hypotheses :

(iii′) f defines an isomorphism of D_+ onto Δ_+;

(iv′) f maps C_0 *onto* Γ_0.

Show then that the extension g is an isomorphism of D on Δ. (It is necessary, in particular, to show that f is simple and, therefore, to show that f takes distinct values at distinct points of C_0; this follows by *reductio ad absurdum* using proposition 4. 2 of chap. III, § 5.)

7. Let a, r be two real numbers such that $r > a > 0$. Find a function $w = f(z)$ which defines an isomorphism of the interior D of Cassini's

oval $|z^2 - a^2| < r^2$ onto the unit disc B, $|w| < 1$, which preserves the axes of symmetry. (Because of the principle of symmetry, it is sufficient to find a function f defining an isomorphism of the right half D^+ defined by

$$|z^2 - a^2| < r^2, \qquad \text{Re}(z) > 0,$$

onto the right half of the unit disc B^+ defined by

$$|w| < 1, \qquad \text{Re}(w) > 0,$$

which takes real values on the real axis and maps the segment iy, $y^2 \leqslant r^2 - a^2$, of the z-plane on the segment iv, $|v| \leqslant 1$, of the w-plane. To do so, consider first the transformation $\zeta = z^2$; then the homographic transformation $Z = \dfrac{\alpha\zeta + \beta}{\gamma\zeta + \delta}$ taking the circle $|\zeta - a^2| = r^2$ into the unit circle $|Z| = 1$ and the segment $a^2 - r^2 \leqslant \text{Re}(\zeta) \leqslant 0$, $\text{Im}(\zeta) = 0$, onto the segment $-1 \leqslant \text{Re}(Z) \leqslant 0$, $\text{Im}(Z) = 0$; and, finally, a suitable branch of the function $w = Z^{1/2}$.)

8. Consider the function $f(z)$ defined in the upper half-plane P^+, $\text{Im}(z) > 0$, by the integral

$$u = f(z) = \int_0^z \frac{dt}{\sqrt{(1-t^2)(1-k^2t^2)}},$$

taken along a path in P^+ joining 0 to z, where k is a real constant such that $0 < k < 1$; we take the simple branch of the integrand which takes the value 1 for $t = 0$. Show that $f(z)$ can be extended to a continuous function in the closed half-plane $\text{Im}(z) \geqslant 0$. Put

$$K = \int_0^1 \frac{dt}{\sqrt{(1-t^2)(1-k^2t^2)}}, \qquad K' = \int_1^{1/k} \frac{dt}{\sqrt{(t^2-1)(1-k^2t^2)}} \quad (t \text{ real}).$$

Show that the function $f(z)$ thus extended defines an isomorphism of the half-plane P^+ onto the (open) rectangle whose vertices are the points $-K$, K, $K + iK'$, $-K + iK'$, and maps the real axis onto the perimeter of this rectangle. Determine which points correspond to the vertices. Show that the inverse transformation $z = F(u)$ can be extended to a meromorphic function which is doubly periodic with periods $4K$, $2iK'$, in the plane of the variable u. Determine its zeros and poles, and deduce that there exists a constant A such that

$$F(u) = A\vartheta_1(u/2K)/\vartheta_0(u/2K),$$

where ϑ_0, ϑ_1 denote the functions considered in exercise 3 of chapter v, with $\tau = iK'/K$.

CHAPTER VII

Holomorphic Systems of Differential Equations

1. Existence and Uniqueness Theorem

1. SETTING THE PROBLEM

Let k be an integer $\geqslant 1$. We are given k analytic functions of $k+1$ real or complex variables :

$$f_i(x, y_1, \ldots, y_k), \qquad 1 \leqslant i \leqslant k.$$

It is assumed that these functions are analytic in some neighbourhood of a point $(a, b_1, \ldots, b_k)$. We consider the system of differential equations

$$\frac{dy_i}{dx} = f_i(x, y_1, \ldots, y_k), \qquad 1 \leqslant i \leqslant k. \tag{1.1}$$

We seek a system of k functions $y_i = \varphi_i(x)$ $(i = 1, \ldots, k)$, which are analytic in some neighbourhood of the point $x = a$ with $\varphi_i(a) = b_i$ and which satisfy the system (1.1). This last condition expresses that the derivatives $\varphi_i'(x)$ satisfy

$$\varphi_i'(x) = f_i(x, \varphi_1(x), \ldots, \varphi_k(x)). \tag{1.2}$$

THEOREM 1. *The above problem has a unique solution.*

This theorem will be proved in the next three sections.

2. CASE WHEN $k = 1$: FORMAL SOLUTION

We have a single unknown function y of the variable x, and the differential equation to be solved is

$$\frac{dy}{dx} = f(x, y). \tag{2.1}$$

The function $f(x, y)$ is given and is analytic in some neighbourhood of the point (a, b). For simplification, we shall assume from now on that $a = 0$ and $b = 0$; we can always reduce it to this case by means of a translation. Let

$$f(x, y) = \sum_{p,\, q \geqslant 0} c_{p,\, q} x^p y^q \tag{2. 2}$$

be the Taylor expansion of f, which, by hypothesis, converges in some neighbourhood of the origin. The unknown function $y = \varphi(x)$ has the Taylor expansion

$$\varphi(x) = \sum_{n \geqslant 1} a_n x^n, \tag{2. 3}$$

whose coefficients a_n are to be determined. The coefficient a_0 is zero since we have required $\varphi(0) = 0$. For the first stage, we shall restrict ourselves to seeking a formal power series (2. 3) which formally satisfies the differential equation (2. 1); in other words, if $\varphi'(x)$ denotes the formal derivative of the formal series $\varphi(x)$, we must have

$$\varphi'(x) = f(x, \varphi(x)), \tag{2. 4}$$

where the right hand side is obtained by substituting the formal series $\varphi(x)$ (without constant term) for y.

PROPOSITION 2. 1. *Given the formal series* (2. 2), *there exists a unique formal series* (2. 3) *which satisfies* (2. 4).

We shall now prove this proposition; we shall not, until later on, consider the question of whether the series (2. 3) thus obtained is indeed convergent in a neighbourhood of 0.

Let us identify the formal series in x on the two sides of (2. 4); by equating the coefficients of x^n in the two sides, we obtain

$$(n + 1)a_{n+1} = P_{n+1}(a_1, \ldots, a_n; c_{p,\, q}), \tag{2. 5}$$

where P_{n+1} is a polynomial in the coefficients $a_1, \ldots, a_n$ and a finite number of the coefficients $c_{p,\, q}$ and the coefficients of the polynomial are integers $\geqslant 0$. It is not worth while giving this polynomial more precisely. For example,

$$a_1 = c_{0,\, 0}, \qquad 2a_2 = c_{1,\, 0} + c_{0,\, 1} a_1.$$

From relations (2. 5), we deduce, by induction on n,

$$a_n = Q_n(c_{p,\, q}), \tag{2. 6}$$

where the Q_n are polynomials, with rational coefficients $\geqslant 0$, in the variables $c_{p,\, q}$ (each polynomial Q_n depends only on a finite number of these variables). It should be carefully noted that the polynomials Q_n are defined once and for all.

The first two polynomials Q_n are

$$Q_1 = c_{0,0}, \qquad Q_2 = \frac{1}{2}(c_{1,0} + c_{0,1}c_{0,0}).$$

Since relations (2. 6) are necessary and sufficient for the formal relation (2. 4) to be satisfied, proposition 2. 1 is proved.

3. Case when $k = 1$: convergence

We now assume that the power series on the right hand side of (2. 2) is convergent in some neighbourhood of (0, 0). We propose to prove that the power series (2. 3), whose coefficients are defined by the formulae (2. 6), then has non-zero radius of convergence. To do so, we shall use the so-called *majorant series* method.

Definition. A formal power series

$$\text{(3. 1)} \qquad F(x, y) = \sum_{p,\, q \geqslant 0} C_{p,q} x^p y^q$$

is called a *majorant series* of the series (2. 2) if the coefficients $C_{p,q}$ are $\geqslant 0$ and satisfy the inequalities

$$|c_{p,q}| \leqslant C_{p,q}.$$

We define similarly a majorant series

$$\text{(3. 2)} \qquad \Phi(x) = \sum_{n \geqslant 1} A_n x^n$$

of the formal series (2. 3).

Proposition 3. 1. *Let $F(x, y)$ be a majorant series of the series $f(x, y)$. Let $\Phi(x)$ be the formal series without constant term, which is the unique formal solution of the differential equation*

$$\text{(3. 3)} \qquad \frac{dy}{dx} = F(x, y).$$

Then, Φ is a majorant series of φ.

Proof. The coefficients A_n of the series Φ are given, as we have seen, by the formulae

$$\text{(3. 4)} \qquad A_n = Q_n(C_{p,q});$$

Since the polynomials Q_n have coefficients $\geqslant 0$, the inequalities $|c_{p,q}| \leqslant C_{p,q}$ and the classical inequalities for the absolute values of a sum or a product show immediately that $|a_n| \leqslant A_n$. This proves the proposition.

To show that the series (2. 3) has non-zero radius of convergence when the given series (2. 2) converges in some neighbourhood of the origin, we shall proceed as follows : we shall choose a majorant series F of the series f, then explicitly calculate the formal solution Φ of the differential equation (3. 3), and verify directly that the series Φ has radius of convergence $\neq 0$. Since the radius of convergence of the series φ is at least equal the radius of convergence of the majorant series Φ, it will follow, as required, that the radius of convergence of φ is $\neq 0$, and theorem 1 (of no. 1) will be completely proved in the case when $k = 1$.

By hypothesis, the series (2. 2) converges in a neighbourhood of the closed set

$$|z| \leqslant r, \qquad |y| \leqslant r \tag{3. 5}$$

where r is number > 0. Let M be the least upper bound of $|f(x, y)|$ on the set (3. 5). By the Cauchy inequalities (chapter IV, § 5, formula (4. 2)) we have

$$|c_{p,q}| \leqslant \frac{M}{r^{p+q}}. \tag{3. 6}$$

Put

$$C_{p,q} = \frac{M}{r^{p+q}}; \tag{3. 7}$$

the $C_{p,q}$ are coefficients of a power series $F(x, y)$ which is a majorant series of $f(x, y)$. The sum of the series F, which is a double geometric series, can be calculated immediately :

$$F(x, y) = \frac{M}{\left(1 - \frac{x}{r}\right)\left(1 - \frac{y}{r}\right)} \quad \text{for} \quad |x| < r, \ |y| < r. \tag{3. 8}$$

The differential equation (3. 3) is separable :

$$\left(1 - \frac{y}{r}\right) dy = \frac{M\,dx}{1 - \frac{x}{r}}. \tag{3. 9}$$

We integrate it directly : the solution y which vanishes when $x = 0$ is given by the relation

$$\left(1 - \frac{y}{r}\right)^2 - 1 = 2M \log\left(1 - \frac{x}{r}\right), \qquad |x| < r, \tag{3. 10}$$

where we take the branch of the logarithm on the right hand side which is zero when $x = 0$. We obtain from (3. 10)

$$y = r\left(1 - \sqrt{1 + 2M \log\left(1 - \frac{x}{r}\right)}\right), \tag{3. 11}$$

where we take the branch of the root on the right hand side which is equal to 1 for $x = 0$. The right hand side of (3. 11) is exactly the function $\Phi(x)$, which is the solution of the differential equation (3. 3) in a neighbourhood of $x = 0$. Moreover, it is clear that this function is analytic in a neighbourhood of $x = 0$ and that its power series expansion thus has radius of convergence $\neq 0$. In fact, it is easy to see that the radius of convergence of the right hand side of (3. 11) is equal to

$$r\left(1 - e^{-\frac{1}{2M}}\right).$$

The proof of theorem 1 is thus completed in the case when $k = 1$.

4. General case, any k

Let us return to the problem set in no. 1. We suppose that $a, b_1, \ldots, b_k$ are zero; it can always be reduced to this case by a translation. The k analytic functions $f_i(x, y_1, \ldots, y_k)$ have Taylor expansions

$$(4.\ 1)\qquad f_i(x, y_1, \ldots, y_k) = \sum_{p, q_1, \ldots, q_k \geqslant 0} c^{(i)}_{p, q_1, \ldots, q_k} x^p y_1^{q_1} \ldots y_k^{q_k}.$$

Given these series, we propose to determine power series

$$(4.\ 2)\qquad \varphi_i(x) = \sum_{n \geqslant 1} a_n^{(i)} x^n$$

which converge in some neighbourhood of 0 and satisfy relations (1. 2). As in the case $k = 1$, we shall argue in two stages: we shall first solve the equations (1. 2) formally; then we shall prove that the power series obtained converge.

The formal solution to equations (1. 2) is unique: we need only write, for each i,

$$\sum_n (n + 1) a^{(i)}_{n+1} x^n = \sum c^{(i)}_{p, q_1, \ldots q_k} x^p \left(\sum_{r_1} a^{(1)}_{r_1} x^{r_1}\right)^{q_1} \ldots$$

This gives, for each i,

$$(4.\ 3)\qquad a_n^{(i)} = Q_n^{(i)}(c^{(j)}_{p, q_1, \ldots, q_k}),$$

where the $Q_n^{(i)}$ are polynomials with rational coefficients $\geqslant 0$ and where each polynomial only depends on a finite number of the variables $c^{(j)}_{p, q_1, \ldots, q_k}$ (the index j taking the values $1, \ldots, k$).

We shall now show that, if the power series (4. 1) converge in a neighbourhood of the origin, the series (4. 2) whose coefficients are given by (4. 3) have non-zero radius of convergence. To do so, we shall replace each series f_i by a majorant series F_i. Let $(\Phi_1, \ldots, \Phi_k)$ be the unique formal solution of the 'majorant system'

$$\frac{dy_i}{dx} = F_i(x, y_1, \ldots, y_k). \tag{4.4}$$

Then, for each i, Φ_i is a majorant of φ_i (the proof is similar to that of proposition 3.1). We need only determine the F_i and the Φ_i explicitly.

By hypothesis, the series (4.1) converge in a neighbourhood of some closed polydisc

$$|x| \leqslant r, \ |y_i| \leqslant r \qquad \text{for} \qquad 1 \leqslant i \leqslant k, \tag{4.5}$$

and their absolute values $|f_i|$ are bounded above by a number M in this polydisc. We deduce that

$$C^{(i)}_{p, q_1, \ldots, q_k} = \frac{M}{r^{p+q_1+\cdots+q_k}} \tag{4.6}$$

are the coefficients of some majorant series F_i. The system of differential equations (4.4) can then be written

$$\frac{dy_i}{dx} = \frac{M}{1 - \frac{x}{r}} \left(\prod_{i=1}^{k} \frac{1}{1 - \frac{y_i}{r}} \right). \tag{4.7}$$

Let $y_i = \Phi(x)$ be unique formal solution of the system of differential equations (4.7). We show that each series Φ_i is equal to a fixed series Φ. For, let $y = \Phi(x)$ be the unique formal solution of the differential equation

$$\left(1 - \frac{y}{r}\right)^k \frac{dy}{dx} = \frac{M}{1 - \frac{x}{r}}. \tag{4.8}$$

It is clear that, if we put $y_i = \Phi(x)$ for all i, we indeed obtain a formal solution of (4.7); this proves the assertion.

Finally, all that is left to be proved is that the formal series $y = \Phi(x)$ which is the solution of the equation (4.8), has non-zero radius of convergence. However the differential equation (4.8) integrates directly: the solution which vanishes for $x = 0$ is

$$y = r \left\{ 1 - \left[1 + (k+1) M \log \left(1 - \frac{x}{r} \right) \right]^{\frac{1}{k+1}} \right\}, \tag{4.9}$$

and the right hand side of (4.9) is certainly an analytic function $\Phi(x)$ in a neighbourhood of $x = 0$; its radius of convergence is then $\neq 0$ as required. In fact, its radius of convergence is equal to

$$r\left(1 - e^{-\frac{1}{(k+1)M}}\right).$$

2. Dependence on Parameters and Initial Conditions

1. Dependence on parameters

We now suppose that the analytic functions f_i which occur on the right hand side of the differential system (1. 1) of § 1 depend analytically on the parameters $t_1, \ldots, t_j$. To be precise, suppose that we are given k functions

$$f_i(x, y_1, \ldots, y_k;\ t_1, \ldots, t_j)$$

which are analytic in the $k+j+1$ variables in some neighbourhood of the origin. For each system $(t_1, \ldots, t_j)$ sufficiently near to $(0, \ldots, 0)$ the differential system

$$\frac{dy_i}{dx} = f_i(x, y_1, \ldots, y_k;\ t_1, \ldots, t_j), \quad (1 \leqslant i \leqslant k) \tag{1. 1}$$

has a unique analytic solution $y_i = \varphi_i(x)$ which is zero for $x = 0$. The functions $\varphi_i(x)$ depend, of course, on the values given to $t_1, \ldots, t_j$. We write

$$y_i = \varphi_i(x;\ t_1, \ldots, t_j), \qquad (1 \leqslant i \leqslant k) \tag{1. 2}$$

for the solution of (1. 1) such that $\varphi_i(0;\ t_1, \ldots, t_k) = 0$ for $i = 1, \ldots, k$.

THEOREM 2. *With the above hypotheses, the functions $\varphi(x_i, t_1, \ldots, t_j)$ are analytic functions of the $j+1$ variables $x, t_1, \ldots, t_j$ in some neighbourhood of the origin $(0, 0, \ldots, 0)$.*

To keep the notation simple, we shall confine ourselves to proving this theorem in the case when $k = 1, j = 1$. Thus we have

$$f(x, y;\ t) = \sum_{p,\, q \geqslant 0} c_{p,\,q}(t) x^p y^q, \tag{1. 3}$$

where the coefficients $c_{p,\,q}(t)$ are themselves power series in t:

$$c_{p,\,q}(t) = \sum_{r \geqslant 0} c_{p,\,q,\,r} t^r. \tag{1. 4}$$

The unique formal solution of the differential equation

$$\frac{dy}{dx} = f(x, y; t) \tag{1. 5}$$

has its coefficients a_n given by the formulae (2. 6) of § 1; thus each a_n is itself a formal series in t. Hence, the formal solution of equation (1. 5) is a formal series

$$(1.6) \qquad y = \varphi(x, t)$$

in the two variables x and t.

To prove theorem 2, it is sufficient to show that the formal series (1. 6) converges whenever x and t are sufficiently small. We do so by using the method of majorant series again. The series (1.3) is, by hypothesis, convergent in a neighbourhood of some closed polydisc

$$(1.7) \qquad |x| \leqslant r, \quad |y| \leqslant r, \quad |t| \leqslant r \quad \text{with} \quad r > 0.$$

Therefore, it has a majorant series $F(x, y, t)$ of the form

$$(1.8) \qquad F(x, y, t) = \frac{M}{\left(1 - \frac{x}{r}\right)\left(1 - \frac{y}{r}\right)\left(1 - \frac{t}{r}\right)}.$$

We notice that this majorant series is deduced from that considered in no 3 of § 1 by replacing the number M by $M\Big/\left(1 - \frac{t}{r}\right)$ on the right hand side of formula (3. 8) of § 1. Thus the solution $y = \Phi(x, t)$ of the majorant differential equation $\frac{dy}{dx} = F(x, y, t)$ is given by the relation

$$(1.9) \qquad \frac{\Phi(x, t)}{r} = 1 - \left[1 + \frac{2M}{1 - \frac{t}{r}} \log\left(1 - \frac{x}{r}\right)\right]^{1/2}.$$

Moreover, it is clear that the right hand side of (1. 9) is an analytic function of the variables x and t in some neighbourhood of the origin $x = 0$, $t = 0$. This completes the proof of theorem 2.

2. Dependence on initial conditions

Let us consider a system of differential equations

$$(2.1) \qquad \frac{dy_i}{dx} = f_i(x, y_1, \ldots, y_k)$$

which, for simplicity, does not depend on any parameters. The given functions $f_i(x, y_1, \ldots, y_k)$ are again assumed to be analytic in some neighbourhood of the origin. If the point $(b_1, \ldots, b_k)$ is near enough to the origin, the functions f_i are also analytic in some neighbourhood of the point $(0, b_1, \ldots, b_k)$. We can then apply the existence and uniqueness theorem (theorem 1 of § 1): there exists a unique solution $y_i = \varphi_i(x)$ of the system of differential equations (2. 1) which is analytic in a neighbourhood of $x = 0$ and is such that $\varphi_i(0) = b_i$. The functions $\varphi_i(x)$ obviously depend on the initial values $b_1, ..., b_k$; we write them $\varphi_i(x, b_1, ..., b_k)$.

THEOREM 3. *With the above notations, the functions* $\varphi_i(x, b_1, \ldots, b_k)$ *are analytic with respect to the the variables* $x, b_1, \ldots, b_k$ *in some neighbourhood of the origin* $x = 0$, $b_1 = 0, \ldots, b_k = 0$.

In other words, the solution of the differential system (2. 1) depends *analytically* on the initial values b_i of the unknown functions y_i.

Proof. We take, as new unknown functions,

$$z_i = y_i - b_i. \tag{2. 2}$$

They must satisfy the differential system

$$\frac{dz_i}{dx} = f_i(x, z_1 + b_1, \ldots, z_k + b_k), \tag{2. 3}$$

with the initial values $z_i(0) = 0$. The right hand sides of the equations (2. 3) depend analytically on the parameters $b_1, \ldots, b_k$ in some neighbourhood of the origin. By theorem 2, the unique solution of (2. 3) which is zero for $x = 0$ is an analytic function $z_i = \psi_i(x; b_1, \ldots, b_k)$. The solution of (2. 1) such that $y_i = b_i$ for $x = 0$ is given by

$$y_i = b_i + \psi_i(x, b_1, \ldots, b_k)$$

and is, consequently, analytic in $x, b_1, \ldots, b_k$ in some neighbourhood of the origin. This proves theorem 3.

3. Differential Equations of Higher Order.

We shall confine our attention to one example, that of a single differential equation of order k :

$$\frac{d^k y}{dx^k} = f(x, y, y', \ldots, y^{(k-1)}). \tag{3. 1}$$

The given function f is an analytic function of $k + 1$ variables in a neighbourhood of the point $(a, b, b_1, \ldots, b_{k-1})$. We seek a function $y = \varphi(x)$ which is analytic in some neighbourhood of the point $x = a$, is such that $\varphi(a) = b$ and is such that the successive derivatives $\varphi'(x), \ldots, \varphi^{(k-1)}(x)$ take the values $b_1, \ldots, b_{k-1}$, respectively, at the point a, and, finally, such that, for x sufficiently near to a, we have the identity

$$\varphi^{(k)}(x) = f(x, \varphi(x), \varphi'(x), \ldots, \varphi^{k-1}(x)).$$

THEOREM 4. *The above problem has a unique solution.*

Proof. We use a classical method of reducing a differential equation like (3. 1) to a differential system of the first order by introducing new unknown functions. To be precise, let us consider, along with the unknown function $y = \varphi(x)$, the functions

$$y' = \frac{d\varphi}{dx}, \ldots, y^{(k-1)} = \frac{d^{k-1}\varphi}{dx^{k-1}}.$$

The functions $y, y', \ldots, y^{(k-1)}$ must satisfy the system of differential equations

$$(3.2) \qquad \begin{cases} \dfrac{dy}{dx} = y', \\ \dfrac{dy'}{dx} = y'', \\ \cdots\cdots\cdots\cdots\cdots, \\ \dfrac{dy^{(k-2)}}{dx} = y^{(k-1)}, \\ \dfrac{dy^{(k-1)}}{dx} = f(x, y, y', \ldots, y^{(k-1)}). \end{cases}$$

We apply theorem 1 of § 1 to the system (3. 2), and this proves theorem 4.

Exercises

1. Given a linear differential equation of the form

$$(1) \qquad (a_0x + b_0)\,y^{(n)} + (a_1x + b_1)\,y^{(n-1)} + \cdots + (a_nx + b_n)\,y = 0,$$

show that, if $U(z)$ is a continuous functions in an open set D, and γ is a piecewise differentiable path in D with initial point z_0 and final point z_1, then the function $f(x)$ defined by the integral

$$(2) \qquad f(x) = \int_\gamma e^{zx}U(z)\,dz$$

is holomorphic in the whole of the x plane. For $f(x)$ to be a solution of (1), show that it is sufficient that

$$\text{(i)} \qquad [e^{zx}A(z)U(z)]_{z_0}^{z_1} = 0,$$

$$\text{(ii)} \qquad \frac{d}{dz}(A(z)U(z)) = B(z)U(z),$$

where

$$A(z) = a_0 z^n + \cdots + a_n, \qquad B(z) = b_0 z^n + \cdots + b_n.$$

Suppose that $A(z)$ has n distinct zeros $c_1, \ldots, c_n$. Show that we can write

$$\frac{B(z)}{A(z)} = \alpha + \frac{\alpha_1}{z - c_1} + \cdots + \frac{\alpha_n}{z - c_n},$$

where $\alpha, \alpha_1, \ldots, \alpha_n$ are complex constants, and deduce that, if γ_j $(j = 1, \ldots, n)$ denotes a closed differentiable path in D, where $D = \mathbf{C} - \{c_1, \ldots, c_n\}$, starting at a fixed point $z_0 \in D$ and encircling the point c_j once, if $\gamma_{j,k}$ $(1 \leqslant j, k \leqslant n)$ denotes the path defined by describing γ_j in the positive sense, γ_k in the positive sense, then γ_j in the negative sense, and finally γ_k in the negative sense, and if we take the function

$$U(z) = \frac{1}{A(z)} e^{\alpha z}(z - c_1)^{\alpha_1} \ldots (z - c_n)^{\alpha_n} \qquad \text{pour} \quad z \in D,$$

a many-valued function in general, then the integral (2), where we take $\gamma = \gamma_{j,k}$ (so $z_0 = z_1$) defines a solution of (1). Show that, at most, $(n - 1)$ solutions (which are holomorphic in the plane) can be found in this way.

2. Proof of the implicit function theorem (propostion 6. 1, chapter IV of § 5, no. 6) by the method of majorant series (we use the notation in the statement of the proposition in question) : show first that it can be reduced to the case where

$$a_j = b_j = c_k = 0, \quad j = 1, \quad \ldots, \quad n; \qquad k = 1, \quad \ldots, \quad p$$

and that

$$(1) \quad f_j(x_1, \ldots, x_n; z_1, \ldots, z_p) = c_{j1}(z)x_1 + \cdots + c_{jn}(z)x_n + \sum_{\nu_1 + \cdots + \nu_n \geqslant 2} c_{j\nu_1 \ldots \nu_n}(z) x_1^{\nu_1} \ldots x_n^{\nu_n},$$

where the coefficients $c_{jj'}(z)$ and $c_{j\nu_1 \ldots \nu_n}(z)$ are themselves power series in $z_1, \ldots, z_p$ and

$$\det c_{jj'}(z) \neq 0$$

for sufficiently small $x_1, \ldots, x_n; z_1, \ldots, z_p$.
Use Cramer's formula to deduce that the system (6. 1) of chapter IV, § 5, is equivalent to

$$(2) \quad x_j = \gamma_{j1}(z) y_1 + \cdots + \gamma_{jn}(z) y_n + \sum_{\nu_1 + \cdots + \nu_n \geqslant 2} \gamma_{j\nu_1 \ldots \nu_n}(z) x_1^{\nu_1} \ldots x_n^{\nu_n},$$

for $j = 1, \ldots, n$, where the coefficients γ are also power series in $z_1, \ldots, z_p$, and that (2) can then be written

$$(3) \quad x_j = \sum_{\substack{1 \leqslant j' \leqslant n \\ \varkappa_1, \ldots, \varkappa_p \geqslant 0}} \gamma_{jj'; \varkappa_1 \ldots \varkappa_p} y_{j'} z_1^{\varkappa_1} \ldots z_p^{\varkappa_p} + \sum_{\substack{\nu_1 + \cdots + \nu_n \geqslant 2 \\ \varkappa_1, \ldots, \varkappa_p \geqslant 0}} \gamma_{j; \nu_1 \ldots \nu_n; \varkappa_1 \ldots \varkappa_p} x_1^{\nu_1} \ldots x_1^{\nu_n} z_1^{\varkappa_1} \ldots z_p^{\varkappa_p}, \; j = 1, \ldots, n.$$

Show that a necessary and sufficient condition for the n formal series

$$(4) \quad x_j = \sum_{\mu_1 + \cdots + \mu_n + \sigma_1 + \cdots + \sigma_p \geqslant 1} d_{j; \mu_1, \ldots, \mu_n; \sigma_1, \ldots, \sigma_p} y_1^{\mu_1} \ldots y_n^{\mu_r} z_1^{\sigma_1} \ldots z_p^{\sigma_p}, \quad j = 1, \ldots, n,$$

to form a system of formal solutions of (3) is that

$$d_{j; \mu_1, \ldots, \mu_n; \sigma_1, \ldots, \sigma_p} = Q_{j; \mu_1, \ldots, \mu_n; \sigma_1, \ldots, \sigma_p}(\gamma, d),$$

where Q denotes a well-defined polynomial with integer coefficients in the $\gamma_{jj'; \varkappa_1, \ldots, \varkappa_p}$, $\gamma_{j; \nu_1, \ldots, \nu_n; \varkappa'_1, \ldots, \varkappa'_p}$, and the $d_{j; \lambda_1, \ldots, \lambda_n; \tau_1, \ldots, \tau_p}$, the latter occurring only if

$$\lambda_1 + \cdots + \lambda_n + \tau_1 + \cdots + \tau_p < \mu_1 + \cdots + \mu_n + \sigma_1 + \cdots + \sigma_p.$$

Deduce that there is a unique formal solution of (3).

To exhibit the convergence of the series obtained, show that (3) has a majorant series of the following form :

$$X_j = \frac{M}{1 - \frac{Z_1 + \cdots + Z_p}{R}} \left\{ Y_1 + \cdots + Y_n + \frac{1}{1 - \frac{X_1 + \cdots + X_n}{R}} - 1 - \frac{X_1 + \cdots + X_n}{R} \right\},$$

where M, R are real positive constants $\left(\text{note that the power series expansion of } \frac{1}{(1 - T_1) \cdots (1 - T_n)} \text{ is majorized by that of} \right.$

$$\left. \frac{1}{1 - (T_1 + \cdots + T_n)} \right).$$

Show that, consequently, by taking $X_1 = X_2 = \cdots = X_n = X$, a majorant series for (4) is obtained by solving the quadratic equation in X :

$$X = \frac{M}{1 - \frac{Z_1 + \cdots + Z_p}{R}} \left\{ Y_1 + \cdots + Y_n + \frac{1}{1 - nX/R} - 1 - \frac{nX}{R} \right\}$$

(see the proof of proposition 9, 1 of chapter 1, § 2, no. 9).

Some numerical or quantitative answers

CHAPTER I

3. $P_2 = a_2b_1^2,\ P_3 = 2a_2b_1b_2 + a_3b_1^3,$
$P_4 = a_2(2b_1b_3 + b_2^2) + 3a_3b_1b_2 + a_4b_1^4,$
$P_5 = 2a_2(b_1b_4 + b_2b_3) + 3a_3(b_1^2b_3 + b_1b_2^2) + 4a_4b_1^3b_2 + a_5b_1^5.$
$X + \frac{1}{3}X^3 + \frac{2}{15}X^5 + \cdots$

4. *a*) infinity, *b*) 1, *c*) $\inf\left(\frac{1}{a}, \frac{1}{b}\right)$.

6. 1.

14. (ii) $n\pi/a$, n an integer.

CHAPTER III

17. (i) $x = \frac{2\,\mathrm{Re}(z)}{1 + |z|^2},\quad y = \frac{2\,\mathrm{Im}(z)}{1 + |z|^2},\quad u = \frac{|z|^2 - 1}{|z|^2 + 1}.$

20. (i) $(\pi(2n-2)!)/(2^{2n-1}[(n-1)!]^2a^{n-1/2}b^{1/2}),$
(ii) $\pi(b-a),$
(iii) $\pi(e^{-a} - 1/2).$
(iv) $\pi a^n/(1-a^2)$ if $|a| < 1$, $\pi/a^n(a^2-1)$ if $|a| > 1$.

23. (ii) $\pi/(n \sin(\alpha + 1)\pi/n).$

25. (i)
$$\sum_{n\geq 1} \frac{1}{a + bn^2} = \frac{1}{2}\left(\frac{\pi}{\sqrt{ab}} \coth \pi\sqrt{\frac{a}{b}} - \frac{1}{a}\right),$$
$$\sum_{n\geq 1} \frac{n^2}{n^4 + a^4} = \frac{\pi}{2\sqrt{2}\,a}\,\frac{\sinh \pi a\sqrt{2} - \sin \pi a\sqrt{2}}{\cosh \pi a\sqrt{2} - \cos \pi a\sqrt{2}},$$
(ii)
$$\sum_{p\geq 1} \frac{1}{x^2 - p^2} = \frac{1}{2x}\left(\pi \operatorname{ctg} \pi x - \frac{1}{x}\right).$$

CHAPTER V

8. $(-1)^n/n!$

9. $a_6 = a_2^2/3, \qquad a_8 = 3a_2a_4/11.$

CHAPTER VI

7. $w = \rho z/\sqrt{a^2z^2 + \rho^4 - a^4}$ taking the branch of the root which is positive real for real z.

The reference numbers indicate in turn
the chapter
the paragraph
the number (or the exercise)

NOTATIONAL INDEX

A CATALOG OF SELECTED

DOVER BOOKS

IN SCIENCE AND MATHEMATICS

A CATALOG OF SELECTED
DOVER BOOKS
IN SCIENCE AND MATHEMATICS

CATALOG OF DOVER BOOKS

ASYMPTOTIC METHODS IN ANALYSIS, N.G. de Bruijn. An inexpensive, comprehensive guide to asymptotic methods—the pioneering work that teaches by explaining worked examples in detail. Index. 224pp. 5⅜ × 8½. 64221-6 Pa. $6.95

OPTICAL RESONANCE AND TWO-LEVEL ATOMS, L. Allen and J.H. Eberly. Clear, comprehensive introduction to basic principles behind all quantum optical resonance phenomena. 53 illustrations. Preface. Index. 256pp. 5⅜ × 8½. 65533-4 Pa. $7.95

COMPLEX VARIABLES, Francis J. Flanigan. Unusual approach, delaying complex algebra till harmonic functions have been analyzed from real variable viewpoint. Includes problems with answers. 364pp. 5⅜ × 8½. 61388-7 Pa. $8.95

ATOMIC SPECTRA AND ATOMIC STRUCTURE, Gerhard Herzberg. One of best introductions; especially for specialist in other fields. Treatment is physical rather than mathematical. 80 illustrations. 257pp. 5⅜ × 8½. 60115-3 Pa. $6.95

APPLIED COMPLEX VARIABLES, John W. Dettman. Step-by-step coverage of fundamentals of analytic function theory—plus lucid exposition of five important applications: Potential Theory; Ordinary Differential Equations; Fourier Transforms; Laplace Transforms; Asymptotic Expansions. 66 figures. Exercises at chapter ends. 512pp. 5⅜ × 8½. 64670-X Pa. $11.95

ULTRASONIC ABSORPTION: An Introduction to the Theory of Sound Absorption and Dispersion in Gases, Liquids and Solids, A.B. Bhatia. Standard reference in the field provides a clear, systematically organized introductory review of fundamental concepts for advanced graduate students, research workers. Numerous diagrams. Bibliography. 440pp. 5⅜ × 8½. 64917-2 Pa. $11.95

UNBOUNDED LINEAR OPERATORS: Theory and Applications, Seymour Goldberg. Classic presents systematic treatment of the theory of unbounded linear operators in normed linear spaces with applications to differential equations. Bibliography. 199pp. 5⅜ × 8½. 64830-3 Pa. $7.95

LIGHT SCATTERING BY SMALL PARTICLES, H.C. van de Hulst. Comprehensive treatment including full range of useful approximation methods for researchers in chemistry, meteorology and astronomy. 44 illustrations. 470pp. 5⅜ × 8½. 64228-3 Pa. $11.95

CONFORMAL MAPPING ON RIEMANN SURFACES, Harvey Cohn. Lucid, insightful book presents ideal coverage of subject. 334 exercises make book perfect for self-study. 55 figures. 352pp. 5⅜ × 8¼. 64025-6 Pa. $9.95

OPTICKS, Sir Isaac Newton. Newton's own experiments with spectroscopy, colors, lenses, reflection, refraction, etc., in language the layman can follow. Foreword by Albert Einstein. 532pp. 5⅜ × 8½. 60205-2 Pa. $9.95

GENERALIZED INTEGRAL TRANSFORMATIONS, A.H. Zemanian. Graduate-level study of recent generalizations of the Laplace, Mellin, Hankel, K. Weierstrass, convolution and other simple transformations. Bibliography. 320pp. 5⅜ × 8½. 65375-7 Pa. $8.95

ROTARY-WING AERODYNAMICS, W.Z. Stepniewski. Clear, concise text covers aerodynamic phenomena of the rotor and offers guidelines for helicopter performance evaluation. Originally prepared for NASA. 537 figures. 640pp. 6⅛ × 9¼.
64647-5 Pa. $15.95

DIFFERENTIAL GEOMETRY, Heinrich W. Guggenheimer. Local differential geometry as an application of advanced calculus and linear algebra. Curvature, transformation groups, surfaces, more. Exercises. 62 figures. 378pp. 5⅜ × 8½.
63433-7 Pa. $8.95

INTRODUCTION TO SPACE DYNAMICS, William Tyrrell Thomson. Comprehensive, classic introduction to space-flight engineering for advanced undergraduate and graduate students. Includes vector algebra, kinematics, transformation of coordinates. Bibliography. Index. 352pp. 5⅜ × 8½. 65113-4 Pa. $8.95

A SURVEY OF MINIMAL SURFACES, Robert Osserman. Up-to-date, in-depth discussion of the field for advanced students. Corrected and enlarged edition covers new developments. Includes numerous problems. 192pp. 5⅜ × 8½.
64998-9 Pa. $8.95

ANALYTICAL MECHANICS OF GEARS, Earle Buckingham. Indispensable reference for modern gear manufacture covers conjugate gear-tooth action, gear-tooth profiles of various gears, many other topics. 263 figures. 102 tables. 546pp. 5⅜ × 8½. 65712-4 Pa. $14.95

SET THEORY AND LOGIC, Robert R. Stoll. Lucid introduction to unified theory of mathematical concepts. Set theory and logic seen as tools for conceptual understanding of real number system. 496pp. 5⅜ × 8¼. 63829-4 Pa. $12.95

A HISTORY OF MECHANICS, René Dugas. Monumental study of mechanical principles from antiquity to quantum mechanics. Contributions of ancient Greeks, Galileo, Leonardo, Kepler, Lagrange, many others. 671pp. 5⅜ × 8½.
65632-2 Pa. $14.95

FAMOUS PROBLEMS OF GEOMETRY AND HOW TO SOLVE THEM, Benjamin Bold. Squaring the circle, trisecting the angle, duplicating the cube: learn their history, why they are impossible to solve, then solve them yourself. 128pp. 5⅜ × 8½. 24297-8 Pa. $4.95

MECHANICAL VIBRATIONS, J.P. Den Hartog. Classic textbook offers lucid explanations and illustrative models, applying theories of vibrations to a variety of practical industrial engineering problems. Numerous figures. 233 problems, solutions. Appendix. Index. Preface. 436pp. 5⅜ × 8½. 64785-4 Pa. $10.95

CURVATURE AND HOMOLOGY, Samuel I. Goldberg. Thorough treatment of specialized branch of differential geometry. Covers Riemannian manifolds, topology of differentiable manifolds, compact Lie groups, other topics. Exercises. 315pp. 5⅜ × 8½. 64314-X Pa. $9.95

HISTORY OF STRENGTH OF MATERIALS, Stephen P. Timoshenko. Excellent historical survey of the strength of materials with many references to the theories of elasticity and structure. 245 figures. 452pp. 5⅜ × 8½. 61187-6 Pa. $11.95

GEOMETRY OF COMPLEX NUMBERS, Hans Schwerdtfeger. Illuminating, widely praised book on analytic geometry of circles, the Moebius transformation, and two-dimensional non-Euclidean geometries. 200pp. 5⅜ × 8¼. 63830-8 Pa. $8.95

MECHANICS, J.P. Den Hartog. A classic introductory text or refresher. Hundreds of applications and design problems illuminate fundamentals of trusses, loaded beams and cables, etc. 334 answered problems. 462pp. 5⅜ × 8½. 60754-2 Pa. $9.95

TOPOLOGY, John G. Hocking and Gail S. Young. Superb one-year course in classical topology. Topological spaces and functions, point-set topology, much more. Examples and problems. Bibliography. Index. 384pp. 5⅜ × 8¼. 65676-4 Pa. $9.95

STRENGTH OF MATERIALS, J.P. Den Hartog. Full, clear treatment of basic material (tension, torsion, bending, etc.) plus advanced material on engineering methods, applications. 350 answered problems. 323pp. 5⅜ × 8½. 60755-0 Pa. $8.95

ELEMENTARY CONCEPTS OF TOPOLOGY, Paul Alexandroff. Elegant, intuitive approach to topology from set-theoretic topology to Betti groups; how concepts of topology are useful in math and physics. 25 figures. 57pp. 5⅜ × 8½. 60747-X Pa. $3.50

ADVANCED STRENGTH OF MATERIALS, J.P. Den Hartog. Superbly written advanced text covers torsion, rotating disks, membrane stresses in shells, much more. Many problems and answers. 388pp. 5⅜ × 8½. 65407-9 Pa. $9.95

COMPUTABILITY AND UNSOLVABILITY, Martin Davis. Classic graduate-level introduction to theory of computability, usually referred to as theory of recurrent functions. New preface and appendix. 288pp. 5⅜ × 8½. 61471-9 Pa. $7.95

GENERAL CHEMISTRY, Linus Pauling. Revised 3rd edition of classic first-year text by Nobel laureate. Atomic and molecular structure, quantum mechanics, statistical mechanics, thermodynamics correlated with descriptive chemistry. Problems. 992pp. 5⅜ × 8½. 65622-5 Pa. $19.95

AN INTRODUCTION TO MATRICES, SETS AND GROUPS FOR SCIENCE STUDENTS, G. Stephenson. Concise, readable text introduces sets, groups, and most importantly, matrices to undergraduate students of physics, chemistry, and engineering. Problems. 164pp. 5⅜ × 8½. 65077-4 Pa. $6.95

THE HISTORICAL BACKGROUND OF CHEMISTRY, Henry M. Leicester. Evolution of ideas, not individual biography. Concentrates on formulation of a coherent set of chemical laws. 260pp. 5⅜ × 8½. 61053-5 Pa. $6.95

THE PHILOSOPHY OF MATHEMATICS: An Introductory Essay, Stephan Körner. Surveys the views of Plato, Aristotle, Leibniz & Kant concerning propositions and theories of applied and pure mathematics. Introduction. Two appendices. Index. 198pp. 5⅜ × 8½. 25048-2 Pa. $7.95

THE DEVELOPMENT OF MODERN CHEMISTRY, Aaron J. Ihde. Authoritative history of chemistry from ancient Greek theory to 20th-century innovation. Covers major chemists and their discoveries. 209 illustrations. 14 tables. Bibliographies. Indices. Appendices. 851pp. 5⅜ × 8½. 64235-6 Pa. $18.95

DE RE METALLICA, Georgius Agricola. The famous Hoover translation of greatest treatise on technological chemistry, engineering, geology, mining of early modern times (1556). All 289 original woodcuts. 638pp. 6¾ × 11.
60006-8 Pa. $18.95

SOME THEORY OF SAMPLING, William Edwards Deming. Analysis of the problems, theory and design of sampling techniques for social scientists, industrial managers and others who find statistics increasingly important in their work. 61 tables. 90 figures. xvii + 602pp. 5⅜ × 8½. 64684-X Pa. $15.95

THE VARIOUS AND INGENIOUS MACHINES OF AGOSTINO RAMELLI: A Classic Sixteenth-Century Illustrated Treatise on Technology, Agostino Ramelli. One of the most widely known and copied works on machinery in the 16th century. 194 detailed plates of water pumps, grain mills, cranes, more. 608pp. 9 × 12.
28180-9 Pa. $24.95

LINEAR PROGRAMMING AND ECONOMIC ANALYSIS, Robert Dorfman, Paul A. Samuelson and Robert M. Solow. First comprehensive treatment of linear programming in standard economic analysis. Game theory, modern welfare economics, Leontief input-output, more. 525pp. 5⅜ × 8½. 65491-5 Pa. $14.95

ELEMENTARY DECISION THEORY, Herman Chernoff and Lincoln E. Moses. Clear introduction to statistics and statistical theory covers data processing, probability and random variables, testing hypotheses, much more. Exercises. 364pp. 5⅜ × 8½. 65218-1 Pa. $9.95

THE COMPLEAT STRATEGYST: Being a Primer on the Theory of Games of Strategy, J.D. Williams. Highly entertaining classic describes, with many illustrated examples, how to select best strategies in conflict situations. Prefaces. Appendices. 268pp. 5⅜ × 8½. 25101-2 Pa. $7.95

MATHEMATICAL METHODS OF OPERATIONS RESEARCH, Thomas L. Saaty. Classic graduate-level text covers historical background, classical methods of forming models, optimization, game theory, probability, queueing theory, much more. Exercises. Bibliography. 448pp. 5⅜ × 8¼. 65703-5 Pa. $12.95

CONSTRUCTIONS AND COMBINATORIAL PROBLEMS IN DESIGN OF EXPERIMENTS, Damaraju Raghavarao. In-depth reference work examines orthogonal Latin squares, incomplete block designs, tactical configuration, partial geometry, much more. Abundant explanations, examples. 416pp. 5⅜ × 8¼.
65685-3 Pa. $10.95

THE ABSOLUTE DIFFERENTIAL CALCULUS (CALCULUS OF TENSORS), Tullio Levi-Civita. Great 20th-century mathematician's classic work on material necessary for mathematical grasp of theory of relativity. 452pp. 5⅜ × 8½.
63401-9 Pa. $9.95

VECTOR AND TENSOR ANALYSIS WITH APPLICATIONS, A.I. Borisenko and I.E. Tarapov. Concise introduction. Worked-out problems, solutions, exercises. 257pp. 5⅜ × 8¼. 63833-2 Pa. $7.95